Measurement of Atmospheric Emissions

Heikki Torvela

Measurement of Atmospheric Emissions

With 105 Figures

Springer-Verlag London Ltd.

Heikki Torvela
Research Professor
Ecocenter
University of Oulu
P.O. Box 400
90571 Oulu
Finland

British Library Cataloguing in Publication Data
Torvela, Heikki Johannes
  Measurement of Atmospheric Emissions
  I. Title
  363.73
  ISBN 978-1-4471-3484-8   ISBN 978-1-4471-3482-4 (eBook)
  DOI 10.1007/978-1-4471-3482-4

Library of Congress Cataloging-in-Publication Data
Torvela, Heikki.
  Measurement of atmospheric emissions / Heikki Torvela.
      p.   cm.
  Includes bibliographical references and index.
  ISBN 978-1-4471-3484-8
  1. Air–Pollution–Measurement.   I. Title.
TD890.T65   1993
628.5′3′0287–dc20   93-31101

Typeset by T & A Typesetting Services, Rochdale Lancashire

69/3830-543210 Printed on acid-free paper

Preface

Emissions into the atmosphere must be controlled in order to maintain a clean environment for man and nature. These emissions easily cross national borders, and international agreements have been formed to establish standards for their control. The United Nations Framework Convention on Climate Change was signed in Rio in July 1992. This agreement is aimed at stabilising atmospheric concentrations of greenhouse gases at a level that would prevent harmful influence on the climate.

It is important to control emissions released from energy production, industry and traffic, as these emissions predominate. To implement reduction technologies, it is necessary to determine the quality and the concentration of the emissions. This can be realised by advanced measurement and analysis techniques.

The need for emission measurements is increasing due to stricter emission limits and standards. In addition to the emission control related to regulations and reduction technologies, emission measurement techniques have an important role for the development of production processes towards lower levels of emissions. This is the way to permanent reduction of emissions and to sustainable technologies.

In this book, measurement technologies are described starting from a physical and chemical basis, with the hope of providing the reader with a favourable basis for critical analysis and investigation of emission measurement techniques, so as to contribute to their development. This is an important issue, as new demands are being imposed on measurement and analysis techniques by legislation and by general opinion. Increasing application of emission measurement techniques increases awareness and provides a reliable basis for legislation. Emission measurement techniques are strongly developing and form a subject of intense interest. Knowledge about the basic principles of emission measurement technologies helps in the practical application of measurement technologies to the monitoring and control of emissions, as widely and increasingly required today. It also helps selection of the most appropriate instruments for various purposes and circumstances.

Present emission measurement technologies utilise diverse physical and chemical principles. The concentration of sulphur dioxide is generally measured by the ultraviolet fluorescence method, which is based on the measurement of the intensity of ultraviolet light emitted by excited sulphur dioxide molecules. The concentration of nitrogen oxides is often measured using the chemiluminescence method, whereby light produced by a chemical reaction is measured. Absorption of infrared light by gas components is also frequently used as the measurement basis. Present techniques used for the measurement of the concentration of gaseous emissions can be schematically presented as in Fig. 1. Chemical reactions or physical phenomena, the extent of which are dependent on the gas concentrations, are utilised. A quantity dependent on the concentration value of a certain gas is converted into an electrical signal, which indicates the concentration.

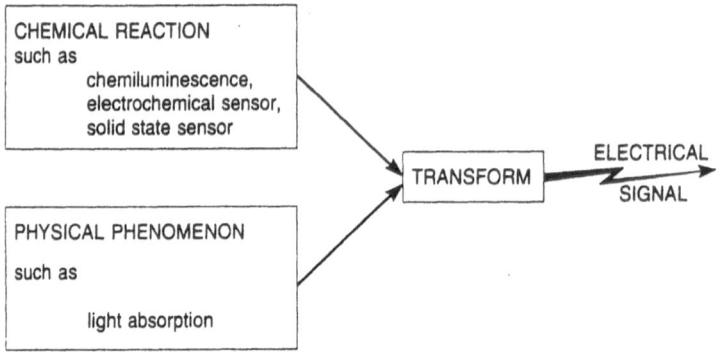

Fig. 1.   Detection of gas concentration and conversion into an electrical signal.

For emission control and analysis, reliable knowledge about the emission gas concentrations as well as the volumetric flow rate of emission gases is required. These requirements can only be met by measurements performed by qualified and experienced measuring teams.

I would like to thank Irja Ruokamo and Juha Huusko for their pleasant co-operation during the preparation of the book.

Oulu, 13 May 1993   Heikki Torvela

Contents

5. Measurement of Gaseous Emissions

6. Optical Spectroscopy in Emission Measurements

7. Other Measurement Technologies

8. Analysis Methods

9. Economics of Emission Measurement and Control

Chapter 1

Origin of Emissions

The quality of air has been a major environmental issue since the 1980s. This is mainly because of the threat of possible global climate change and the "greenhouse effect" caused by constituents of anthropogenic origin in the atmosphere.

The major part of the air contaminants causing pollution in towns, for instance, originates from industrial production, transportation and the production of heating energy and electricity. All these utilise largely the combustion of fuels. The main gaseous products emitted from the combustion processes of fossil fuels are nitrogen (N_2), carbon dioxide (CO_2), oxygen (O_2) and water (H_2O), which are not considered harmful. Of these carbon dioxide and water vapour are so called "greenhouse" gases. There are other gas components, however, formed during the combustion or the oxidation of the fuels. These include:

- Sulphur dioxide (SO_2) and nitrogen oxide (NO) from the oxidation of sulphur and nitrogen contained in the fuel
- Oxidation of a minor part of the nitrogen of air into NO
- Ash particles containing elements such as heavy metals, polycyclic aromatic hydrocarbons (PAHs), and soot
- Carbon monoxide (CO) as a result of incomplete burning.

1.1. Emissions from Energy Production

1.1.1. Sulphur Oxides

Sulphur contained in the fuel can react in the combustor in many ways [1]. Most of it is first oxidised into sulphur dioxide (SO_2), a small portion of which (normally less than 1%) is later converted in the flue gas into sulphur trioxide

(SO_3) and into gaseous sulphuric acid (H_2SO_4). In some cases, e.g. in soda recovery boilers, part of the emissions of sulphur can also be hydrogen sulphide (H_2S) or organic reduced compounds of sulphur.

Emissions of sulphur oxides (SO_x) in the USA, for instance, peaked in 1973 at about 32 million tons per year [2]. By the late 1970s, increased numbers of SO_2 control systems and decreased sulphur contents of fuel oils had reduced emissions significantly. It is estimated that in 1981, SO_x emissions decreased below 25 million tons for the first time in more than a decade (US Environmental Protection Agency data, 1982). About two-thirds of the sulphur oxides in the USA are emitted from coal-fired power plants. Industrial fuel combustion and industrial processes (primarily petroleum refining, H_2SO_4 manufacturing, and smelting of nonferrous metals) account for the most of the remainder of the sulphur oxides.

In 1980 Dutch power stations emitted 210 000 tons of sulphur dioxide [3]. In 1982, more natural gas was made available for electricity generation, and the emissions were reduced by 70%. Further reductions in emissions will obviously be obtained, despite an increase in the use of coal, by shutting down 1000 MW of older coal-fired plants and by the construction of flue gas desulphurisation plants in existing power plants.

1.1.2. Nitrogen Oxides

The formation of the nitrogen oxides in combustion processes is a complicated physicochemical process [1]. Nitrogen oxides can be formed from either the nitrogen contained in the fuel (fuel NO_x) or from the nitrogen contained in air by a thermal reaction (thermal NO_x). In addition to the fuel used, the emissions of nitrogen oxides are thus strongly related to the combustion technology. The efforts towards the reduction of nitrogen oxides are significantly based on the development of the technology of combustion.

The seven oxides of nitrogen that are known to occur are NO, NO_2, NO_3, N_2O, N_2O_3, N_2O_4, and N_2O_5 [2]. Of these, nitric oxide (NO) and nitrogen dioxide (NO_2) are the two most important air pollutants. In the emission gas of an ordinary combustion process, nitrogen oxides are mainly (more than 90%) nitrogen monoxide (NO). Much of the rest is nitrogen dioxide (NO_2), and often in air pollution work "NO_x" refers to NO and NO_2. In some cases, depending on the combustion techniques, nitrous oxide (N_2O) has been noticed to be present in emission gases [4].

Nitrogen oxides are emitted in the USA at a rate of about 20 million tons per year [2]. About 40% of this is emitted from mobile sources. Of the 11–12 tons of nitrogen oxides that originate from stationary sources, about 30% is the result of fuel combustion in large industrial furnaces and 70% is from electric utility furnaces.

As mentioned earlier, the emissions of NO_x vary depending on the type of fuel and the type of firing [2]. Some examples of emission factors (without controls) are presented in Table 1.1.

Table 1.1. Selected NO_x emission factors (without controls) [2]

Type of boiler	Type of fuel	Emissions factor[a] $kg\,NO_x\,Mg^{-1}$ of coal (or as indicated)
Pulverised coal	Anthracite	9
Pulverised coal		
Dry bottom	Bituminous and sub-bituminous	$10.5\,(7.5)^b$
Wet bottom	Bituminous and sub-bituminous	17
Cyclone (crushed coal)	Bituminous and sub-bituminous	18.5
Pulverised coal	Lignite	6–$7\,(4)^b$
Utility boilers		
Tangentially fired	Residual oil	5^c
Vertically fired	Residual oil	12.6^c
All (in general)	Natural gas	8800^d
Industrial boilers		
	Residual oil	6.6^e
	Distillate oil	2.4^c
	Natural gas	2240^d

Source: US Environmental Protection Agency data, 1982

Notes: [a] Total nitrogen oxides are expressed as NO_2.
[b] The values in parentheses are for tangentially fired units.
[c] Emissions are expressed as $kg\,NO_2\,(1000\,l\ oil\ burned)^{-1}$.
[d] Emissions are expressed as $kg\,NO_2\,(10^6\,m^3\ gas\ burned)^{-1}$.
[e] NO_x emissions in industrial boilers depend strongly on the nitrogen content of the residual oil.

The emissions of nitrogen oxides in the early 1980s in Great Britain, for instance, were in the region of 2 million tons per year [5]. In Sweden the yearly emissions over the same period in the region of 250 000 tons. In 1987, the emissions of nitrogen oxides in Finland were calculated to be about 260 000 tons [6]. Of this, 33.6% originated from energy production, 46.7% from road traffic, 11.6% from other transportation and 8% from industry.

1.1.3. Particulate Materials

Particulates are emitted from many different sources, including both combustion and noncombustion processes in industry, mining and construction activities, motor vehicles and refuse incineration [2].

The level of the particle emissions from an energy plant is mainly controlled by the ash content of the fuel, the distribution of the ashes between the bottom and fly ash, and the efficiency of the reduction technology utilised by the plant [1].

The ash content of the solid fuels is high. Coal contains about 12%–16% ash, and peat, for instance, about 3%–6%. The ash content of the heavy fuel oil is in the range of 0.01%–0.03%.

The solid matter in the emission gas of a combustion process consists of three constituents [1]:

1. The inorganic residue of the fuel (ash)
2. The unburnt part of the fuel particles or drops (char)
3. Fine-grained carbon (soot) formed by the incomplete burning of pyrolysis gases

The size of the ash particles is normally in the range 10–100 μm, similar to the size of the char particles. Soot is significantly finer, being generally less than 0.1 μm.

1.1.4. Heavy Metals

Heavy metals are generally considered to include [1] arsenic (As), chromium (Cr), nickel (Ni), vanadium (V), lead (Pb), cadmium (Cd) and mercury (Hg). Beryllium (Be), cobalt (Co), manganese (Mn), molybdenum (Mo), zinc (Zn) and copper (Cu) are also often considered as heavy metals.

Solid fuels always contain heavy metals. Their concentration is high, e.g. in coal. Heavy metals are bound in different compounds in the form of sulphides, for instance. In the combustion, part of the heavy metals will remain in the ash, but some part of the easily vaporised heavy metals can be transported into the atmosphere, especially when the cleaning treatment of the gases is inefficient.

Part of the heavy metals remaining in the ash will be retained in the bottom ashes, whereas another part will be transported by the particulate material of the flue gas into the flue channel [1]. Thereafter they are, for the most part, separated in particle collectors. The emission of heavy metals is formed by only the part penetrating the particle collector. This is normally a very small mass portion of the original particle emission. Typical heavy metals following the ash are manganese and cobalt.

Many metals, especially Hg, Pb, As and Cd, and in some cases also V, will evaporate at least in part under proper conditions. They can thus be carried into flue channels together with flue gases, and be precipitated at the surfaces of particulates, or continue in a partially gaseous state into the atmosphere. Part of the metals precipitated at the surfaces of the particles will be captured by the collectors when collecting particles.

1.1.5. PAH Compounds

Part of the polycyclic aromatic hydrocarbon (PAH) compounds originates from the fuel, and part is formed during the combustion. The formation of PAH compounds is associated with incomplete combustion, similar to the formation of carbon monoxide. Preconditions for the generation of PAH compounds include that the fuel remains in the combustor under reducing conditions for a proper period of time at a temperature less than 900 °C. For example, an optimum temperature for the formation of benzo(a)pyrene is around 700 °C [7]. In addition to the combustion temperature and the stoichiometry, the generation of PAH-compounds is strongly dependent on the mixing of air and fuel. The generation can be reasonably low, if the mixing of air and fuel is efficient, enough air is available, and combustion gases remain in the combustion area for a long enough time.

PAH compounds are normally adsorbed almost completely on particles. Gaseous PAH compounds originating from fossil fuels are adsorbed on fly ash after having cooled down in the flue gas. The most part is fixed on small particles of less than 3 µm.

PAH emissions can be remarkable from small boilers using solid fuels. According to measurements, relative emissions of PAH compounds from small wood-fired ovens is approximately ten times those from ovens of 1 MW. Correspondingly, emissions from this size of boilers of about 1 MW, are again approximately ten times the emissions of the boilers of about 50 MW, as related to fuel energy [1].

1.1.6. Emissions of PCDDs and PCDFs

The major sources of emission of polychlorinated dibenzo-p-dioxins (PCDD) or "dioxins" and the closely related polychlorinated dibenzofurans (PCDF) or "furans" to the environment from combustion sources are generally regarded as municipal waste incinerators [8]. Other sources include industrial and domestic coal fires, vehicle exhausts, hospital incinerators and coal-fired power stations [9]. PCDDs and PCDFs can be formed, if there is chlorine present in the combustion of organic matter [10]. Emissions of PCDD and PCDF can also originate from chemical industries, pulp and paper industry and metal smelters.

A number of theories have been proposed for the formation of PCDD and PCDF during combustion [8]. Their formation route can be a combination of processes, as determined by the prevailing conditions.

PCDD and PCDF occur as trace constituents in the waste, and because of their thermal stability they survive the combustion process [8]. Mass balances have shown that higher concentrations have been found in the emissions than are found in the input.

PCDD and PCDF are produced during the incineration process from precursors such as polychlorinated biphenyls (PCB), chlorinated benzenes and pentachlorophenols [8]. The in-situ synthesis of PCDD and PCDF therefore occurs via re-arrangement, free-radical condensation, dechlorination and other molecular reactions. Laboratory studies have shown that these compounds can be formed from chlorophenols at 280–300 °C, PCBs or chlorobenzenes at 550–650 °C and brominated diphenylethers at 600–800 °C. These precursors may be present in the waste or formed by the combustion of plastics such as PVC.

PCDD and PCDF are also produced as a result of elementary reactions of the appropriate elements, carbon, hydrogen, oxygen and chlorine atoms. This reaction is called a de-novo synthesis of PCDD and PCDF [8]. De-novo synthesis has been cited to take place in the combustion plasma or in the plume after combustion. Also PCDD and PCDF have been shown to form on fly ash that contains residual carbon collected within a combustion system at temperatures in the region of 300–400 °C in the presence of flue gases containing HCl, O_2 and H_2O.

The generalised molecular structure of PCDD and PCDF is tricyclic aromatic compounds containing two (dioxin) or one (furan) oxygen atoms [8]. Each of these structures represents a whole series of discrete compounds having between one and eight chlorine atoms attached to the ring substituting for hydrogen. For example, the tetra isomers contain four chlorine atoms.

There are 75 PCDD isomers and 135 PCDF isomers. The most toxic dioxin is 2378-tetrachlorinated dibenzo-p-dioxin having the four chlorine atoms in the 2378 positions. Assessment of the toxicity of PCDD and PCDF mixtures has led to the development of the Toxic Equivalent (TEQ) scheme, in which the available toxicological and biological data are used to generate a set of weighting factors, each of which expresses the toxicity of a particular PCDD or PCDF in terms of an equivalent amount of the most toxic and most analysed PCDD, 2378 TCDD [8].

1.1.7. Other Emissions

In addition to the examples given above, other gases are released into the atmosphere from the energy production. These gases include carbon monoxide, carbon dioxide and hydrocarbons. Carbon dioxide is well known as a greenhouse gas. The control of carbon dioxide, based on the threat of the climate change, has been discussed in many international meetings. Many countries have made some sort of commitment to reduce or stabilise carbon dioxide and other greenhouse gas emissions, even through technological means for it are limited.

1.2. Emissions from Process Industry

1.2.1. Chemical Industry

Gaseous emissions from oil refineries consist mainly of sulphur compounds, nitrogen oxides and hydrocarbons. In many oil refineries, organic sulphur from oil is removed in desulphurisation units via a catalytic reaction with hydrogen to form hydrogen sulphide (H_2S) [2]. H_2S is often captured (more than 98%) by using the Claus process, whereby part of the H_2S is burned to SO_2, and then the two compounds are combined over a catalyst to oxidise and reduce each other simultaneously. The elemental sulphur is separated in the molten state.

This reduces the emissions of either H_2S or SO_2 from a refinery substantially. In refineries, some particulate material and metals such as vanadium and nickel are emitted from the combustion of heavy fuel oil.

Main emissions from the organic chemical industry include aliphatic, aromatic and chlorinated hydrocarbons and organic sulphur compounds [11].

The inorganic chemical industry produces gaseous emissions such as sulphur oxides, ammonia and nitrogen oxides [11]. Most of the sulphur emissions originate from the production of sulphuric acid and from the combustion processes.

1.2.2. Metal Industry

The most important emissions from the metal industries are of sulphur dioxide and particulates. In the production processes of some nonferrous metals, sulphide concentrates are oxidised raising the concentration of SO_2 in gases emitted to several percent. This gas is generally used to produce sulphuric acid in a sulphuric acid plant following the metal smelters. The total emission of sulphur dioxide is thus efficiently controlled.

1.2.3. Production of Building Materials

The production of concrete, lime and limestone causes mainly particle emissions. From the production processes of insulating materials, some emissions of carbon monoxide, phenol and phenol compounds can result.

1.2.4. Volatile Organic Compounds (VOCs)

Volatile organic compounds (VOCs) are a broad class of substances emitted from diverse sources [12]. The main part of the hydrocarbon emissions originates from industries using solvents, from traffic, oil refineries, the petrol delivery chain and the production of energy. In addition to pure hydrocarbons, VOCs can also include aldehydes, ketones, chlorinated solvents, and so on. Common to most of the volatile organic compounds, in addition to volatility, is their contribution to photochemical reactions initiated and sustained by solar radiation.

1.3. Emissions from Traffic

Today's traffic systems rely on fossil fuels like oil [13]. In the combustion of fossil fuels, emissions are released into the atmosphere. These emissions consist of carbon dioxide, carbon monoxide, oxides of sulphur and nitrogen, hydrocarbons and particles, which can contain substances harmful to health. Of the oxides of nitrogen, for instance, 40%–60% are emitted from mobile sources.

Most of the emissions from traffic into the atmosphere originate from road vehicles [13]. Pollution caused by traffic emissions is most pronounced in densely populated communities, as exhaust gases are released into the atmosphere at low heights, near the objects of exposure. The mixing of exhaust gases with fresh air in these areas is also slow due to high building density.

1.4. Transport of Emissions

Emissions in the atmosphere are carried along with air movements. Depending on the lifetime (Table 1.2) and persistence of emission components in air, their transport effects can range from global and long-range effects to local ones. Transformation of emission gases via atmospheric chemical reactions is one factor influencing the atmospheric dispersion of gas compounds. Other factors include, of course, weather conditions and factors affecting wind patterns.

Table 1.2. Approximate lifetimes of some gas compounds in air [14–16]

Carbon dioxide (CO_2)	tens of years
Carbon monoxide (CO)	2–6 months
Nitrogen dioxide (NO_2)	8–10 days
Nitrous oxide (N_2O)	150 years
Sulphur dioxide (SO_2)	2 days
Ammonia (NH_3)	5 days
Hydrogen sulphide (H_2S)	0.5 day
Water vapour (H_2O)	10 days
Methane (CH_4)	10 years
Chlorofluorocarbons (CFCs)	60–100 years
Tropospheric ozone (O_3)	1 day

1.4.1. Global Air Circulation Patterns

A simplified air circulation pattern of the earth can be visualised based on air pressure differences and the rotation of the globe. The forces determining the wind direction and velocity are [2]:

- Pressure gradient force
- Deflective force or Coriolis force caused by the rotation of the earth
- Friction force
- Centrifugal force

The driving force for air movements is the existence of pressure gradients. The other forces act to modify the direction and velocity of flow produced by the gradient force.

There are many factors which affect the wind pattern at the surface of the earth and make it to differ from that originally predicted [2]. These include:

- Topography
- Diurnal variation and seasonal variation in surface heating
- Variation of surface heating owing to the presence of ground cover
- Proximity to large bodies of water

1.4.2. Local Circulation Effects

The origin of local circulations is unequal surface heating and cooling [2]. The land–sea breeze occurs, where sea and land meet. In the daytime, the land area is heated rapidly, which in turn heats the air just above it. The water temperature remains relatively constant. The air over the heated land surface rises, producing a low pressure relative to the pressure over the water. The resulting pressure gradient produces a surface flow off the water toward the land. At night, the reverse flow pattern will develop.

Mountain–valley winds are generated when, in the daytime, the air adjacent to the mountain slope heats rapidly and rises [2]. This air then settles over the cooler valley, producing an up slope wind during the day. At night, the cooler air on the mountain slope flows down into the valley.

1.4.3. Wind Roses

Average wind data for some areas can be presented graphically using a wind rose, which expresses time fractions in percentages of winds from different directions. Wind velocity ranges can also be expressed in wind roses.

1.4.4. Stability and Vertical Mixing

The plume from a tall stack can first rise vertically and then abruptly spread out horizontally [2]. This indicates, that under certain conditions, the lower layer of the atmosphere can resist vertical mixing. The resistance to vertical mixing is referred to as stability. The phenomenon is based on an abrupt change in the vertical temperature profile occuring under certain conditions.

1.4.5. Dispersion Modelling

The accumulation of pollutants in any localised region is dependent on emission rates, dispersion rates, and generation or destruction rates (by chemical reactions) [2]. The dispersion of pollutants is almost entirely dependent upon local meteorological conditions, such as wind speed and direction, and atmospheric stability.

In dispersion model calculations, the plume must be considered on a time-average rather than an instantaneous basis. The time-averaged pollutant concentration at a given distance, x_0, downwind from the source has been shown to be normally distributed in the $\pm y$ direction. A similar spreading of the plume occurs in the vertical direction, resulting in another normal distribution of pollutant concentration. The distribution of pollutant is termed binormal. An approach generally used for the dispersion modelling calculations is based on the statistical nature of the dispersion process [2]. This model is usually referred to as the Gaussian dispersion equation, and is found in many texts on distribution model calculations.

1.4.6. European Monitoring and Evaluation Programme

The co-operative programme for monitoring and evaluation of the long range transboundary air pollutants in Europe, generally known as EMEP (the European Monitoring and Evaluation Programme), was started in 1977 and adopted in Geneva in September 1984 [17]. It is based at two centres, EMEP-West in Oslo, Norway and EMEP-East in Moscow, Russia, and is funded by 27 member countries. The programme is carried out under the United Nations Economic Commission for Europe (UNECE) Long Range Transboundary Air Pollution Control Convention.

The calculation of the airborne transport of the various pollutants between the countries is one of the main responsibilities of the EMEP. There are about 90 measuring stations in 25 countries regularly sending in their measurements of the concentrations and depositions of pollutants. The annual national total emissions of the various pollutants are also submitted to the EMEP. In 1990, measurements included sulphur, nitrogen oxides, ammonia, volatile organic compounds (VOCs) and photochemical oxidants [17].

1.4.7. International Agreements on Air Pollution Control

The Convention on Long-Range Transboundary Air Pollution was signed in Geneva in 1979 by 33 countries and was also signed and ratified by the EEC by 1989 [17]. It is a framework convention under which states recognise the transnational problems of air pollution, and accept their general responsibility to move towards a solution to these problems. The text requires countries gradually to reduce and prevent air pollution and to exchange information on significant changes in pollution levels, on control technologies and approaches, and to take account of the transboundary effects in authorising any major new pollutant sources. The Convention came into force in March 1983 after ratification by 24 of the signatories [17].

The Executive Body of the Convention first met in June 1983 and agreed the necessity of developing abatement programmes to reduce SO_2 emissions by the mid-1990s [17]. Over the following two years, this led to proposals for a legally binding SO_2 protocol to the Convention, committing signatories to reduce their SO_2 emissions by 30% from 1980 levels by 1993 (Helsinki Protocol). Eleven countries of the signatories have announced that they will reduce SO_2 emissions by 50% by 1995.

An equally binding protocol, the Sofia Protocol, freezing NO_x emissions at 1987 levels, or at that of any earlier year before 1995, was signed by 23 countries in 1988. This protocol came into force on 14 February 1991 after ratification by 16 of its signatory countries. A further agreement, called the NO_x declaration, to reduce NO_x emissions by 30% compared with 1987 levels by 1988 was signed by 12 countries [17].

A general agreement for the protection of the ozone layer was signed in Vienna in 1985. Associated with this agreement, the protocol on the compounds reducing the ozone layer was signed in Montreal in 1987.

The United Nations Framework Convention on Climate Change was signed in Rio in July 1992. This agreement is aimed at stabilising atmospheric concentrations of greenhouse gases at a level that would prevent harmful influence on the climate.

1.5. Legislation and Emission Reduction Policies

The development of the environmental legislation, based on concern for the quality of the surrounding atmosphere, is setting increasing requirements on emission control strategies and technologies. In the USA, for instance, the Clean Air Act Amendments of 1990 require electric utilities to reduce SO_2 emissions by approximately 10 million tons per year by 2000 [18]. The Electric Power Research Institute (EPRI) has set a goal to develop and demonstrate SO_2 control technology which can achieve 50%–70% SO_2 removal at costs of $500–$1000 ton^{-1} SO_2 [19]. NO_x emission limits are also becoming stricter. Particulate emission limits put forward by New Source Performance Standards are lower than the earlier ones setting new requirements on the performance of electrostatic precipitators and baghouses.

In Germany, coal- and oil-fired utility boilers totalling more than 45 000 megawatts of electricity (MW_e) have been retrofitted with flue gas desulphurisation (FGD). Of these, boilers to a total of more than 34 000 MW_e have been provided with selective catalytic reduction (SCR) of the oxides of nitrogen [20]. In Germany, the new TA Luft replaces the 1974 standards for all emissions except power plants with over 50 MW thermal capacity, which are covered in the large power plant ordinance [21].

Additional restrictions were set for NO_x by a resolution of the State Ministers for Environmental Affairs (UMK) in 1984. Regulations were made tighter by local agreements, e.g. in the State of Northrine-Westfalia.

In the UK, flue gas desulphurisation will be required on all new boilers and some retrofits (totalling 20 000 MW) in the next few years [21]. According to the EEC policy on the reduction of emissions, the Central Electricity Generating Board is developing processes for the reduction of sulphur oxides as well as nitrogen oxides.

There are about 1500 wet process FGD plants in Japan totalling about 50 000 MW [22]. These processes achieve 90%–98% SO_2 removal efficiency. Dry and semi-dry injection is applied at more than 100 municipal solid waste incinerators to reduce SO_x and HCl emissions by 30%–90%. More than 250 selective catalytic reduction plants are in operation, mainly at utility boilers in Japan to reduce NO_x emissions by 50%–85%. Selective noncatalytic reduction is used at about 50 smaller gas sources including municipal solid waste incinerators. Combined systems, to remove both NO_x and SO_2 using SCR followed by FGD, are popular in Japan.

In 1980 Dutch power stations emitted 210 000 tons of sulphur dioxide [3]. In 1982 more natural gas was made available for electricity generation, and the emissions were reduced by 70%. Further reductions in emissions will be obtained, despite an increase in the use of coal, by shutting down older coal-fired plants to a total of 1000 MW and by the construction of flue gas desulphurisation plants in existing power plants. The reduction of the emissions of nitrogen oxides is pursued in the framework of the NO_x Abatement Programme of the co-operating power production companies. The emission of particulates is reduced by the use of high efficiency electrostatic precipitators. Power production companies are obliged to measure the emissions [23]. Combustion plants over 300 MW must measure continuously the emissions of SO_2, NO_x and particulates. The Dutch electricity supply companies have set up a company with the task to market the slag, fly ash and FRG by-products and to develop new applications.

Italy, as one of the co-signers of the Helsinki protocol, has undertaken to reduce its overall SO_2 emissions 30% by 1993 compared with the 1980 level [21]. According to the NO_x protocol signed in Sofia, and the later Declaration, Italy has committed to reduce NO_x emissions by 30% compared with any year between 1980 and 1986. After the nuclear moratorium decided in Italy, ENEL (Ente Nazionale per L'Energia) Planning Department has been concerned with the emission control [24].

For the control of sulphur dioxide, ENEL is planning to adopt a suitable mix of design and fuel options. In coal-fired plants a complete design option is used. A type of wet limestone-gypsum process will be installed in all coal-fired plants. In oil- or gas-fired plants a mix of design and fuel choice options will be adopted [24].

For the control of nitrogen oxides, two concurrent measures will be taken, namely the installation of

- Low-NO_x burners and/or recourse to various "primary measures" to be implemented in the boiler (overfiring, stage combustion, reburning)
- Selective catalytic reduction (SCR) system

In addition to these planned actions, ENEL is carrying out extensive research on:

- FGD using "regenerable" processes
- Selective noncatalytic reduction, for plants having boilers originally designed for coal firing but using oil

Finland, as well as the other Nordic countries, has signed both the SO_2 protocol in Helsinki and the NO_x protocol in Sofia. Finland has already reduced SO_2 emissions by about 50%. According to the decision of the Finnish Government, SO_2 emissions are intended to be reduced by 80% by 2000 from the level of 1980.

The development of the emission reduction technologies will also be affected by the concepts of emission credits or emission trading. In the USA, for instance, it may be advantageous to remove more SO_2 and NO_x than required [25]. In Sweden, a sum per kilogram of NO_2 must be paid, which corresponds to the cost of an SCR unit [26].

One problem is that piecemeal legislation creates piecemeal control technology [25]. If legislation is not made to encourage an integrated approach to pollution control, and if the industry is not structured so that power generators have incentives to select a riskier, but potentially less expensive technology, then techniques like simultaneous SO_2/NO_x processes have little future [25].

1.5.1. Regulations and Standards

In the USA, the two types of standards in the federal legislation are the National Ambient Air Quality Standards (NAAQSs) and the National Source Performance Standards (NSPSs) [2]. Ambient Air Quality Standards are written in terms of concentration ($\mu g\, m^{-3}$ or ppm (parts per million)), whereas Source Performance Standards are written in terms of mass emission per unit of time or unit of production ($g\, min^{-1}$ or kg of pollutant per ton of product produced). National Ambient Air Quality Standards have been set by the Environmental Protection Agency (EPA) based on two criteria: the primary standards have been established to protect the public health, whereas the secondary standards have been established to protect the public well-being.

National Ambient Air Quality Standards have been established for six primary air pollutants and one secondary pollutant (ozone, formed by chemical reactions in the atmosphere) [2]. The five major primary pollutants are particulates, sulphur oxides (SO_x), nitrogen oxides (NO_x), volatile organic compounds (VOCs), and carbon monoxide (CO). The sixth primary pollutant is particulate lead.

1.5.2. Emission Standards and Measurement Basis

Emission standards applicable to coal combustion, for instance, are generally set for the three major pollutants, which are

- Particulate material (or dust)
- Sulphur dioxide (SO_2)
- Nitrogen oxides (NO_x)

For other pollutants, emission standards are set in some countries. In Germany, standards for coal-fired plants over 50 MW_t (thermal) also cover CO, HCl and HF. In the UK, the concentration of ammonia in emissions is required to be kept to the lowest practicable value [17]. This applies also to similar compounds used as additives which are injected into the gas stream. In the UK, controls over emissions of chlorine compounds (halogens); metals, metalloids and their compounds; organic compounds and phosphorus and its compounds have also been introduced. The Environmental Protection Act dictates the minimisation of the impacts and restriction of these emissions on the environment and requires this in the authorisation. In the USA, the Clean Air Act Amendments (CAAA) require limits to be set for 189 hazardous (to human health or the environment) or toxic pollutants emitted by stationary sources other than utility power plants [17]. The best commercial control technology developed will be used to control these pollutants.

In Germany, standards are based on measurement of gas volumes at 0 °C (273 K), 101.3 kPa on dry flue gases. The reference oxygen content in flue gas for solid fuels varies as follows under [17]:

- TA Luft (Technical Instruction for Air Pollution Control, plants
  below 50 MW_t):   7%
- GRAVO (Ordinance on Large Combustion Plants)
  FBC (Fluidised Bed Combustion) and grate firing: 7%
  pulverised fuel (dry bottom):   6%
  pulverised fuel (wet bottom):   5%

Emission standards must be met on a daily average basis, with 97% of half hourly values less than or equal to 1.2 × standard, 100% of half hourly values less than or equal to 2 × standard.

In France, emission standards are given in mg m^{-3} STP (0 °C (273 K), 101.3 kPa), at an oxygen content of 6% on dry flue gas [17]. In Japan, plant sizes are measured in m^3 h^{-1}, but are converted to MW_t to facilitate comparison with plants in other countries (plant size of 1000 m^3 emission gas volume is equivalent to approximately 0.8 MW_t). Emission standards are based on gas volumes measured at standard temperature and pressure (273 K, 101.3 kPa) and 6% oxygen on dry flue gases for coal-fired boilers, 10% for cement kilns and 7% for coke ovens [17].

In the UK, plant sizes are expressed on the basis of MW thermal input or net heat value of fuels burned at maximum continuous rating. Emission limits are expressed as $mg\,m^{-3}$ for particulates, sulphur and NO_x, based on gas volumes at $0\,°C$ (273 K), 101.3 kPa, and 6% O_2 on dry flue gas. The SO_2 and NO_x emission limits are assessed on a monthly calendar average basis where in a calendar year no more than 5% of all 48 h averages should exceed 110% of the set limits; particulate limits are set at a 2 h average continuous operation [17].

Plant sizes are based on MW maximum thermal input capacity in the USA. Emission standards are based on the energy input as calorific or gross heat values and must be met on a 30-day rolling average. Conversions are assuming flue gas volume $350\,m^3\,GJ^{-1}$ (gross heat value) at standard temperature and pressure (273 K, 101.3 kPa) on dry flue gases [17].

1.6. Effects of Emissions

1.6.1. Deposition of Sulphur

The deposition of sulphur is considered to cause [5]

- Acidification in soil, lakes and water systems
- Dissolution of nutrients and consequently damage to forests over time
- Damage to materials, buildings and cultural monuments by corrosion
- Gas damage in trees and other plants
- Harm to human health

1.6.2. Deposition of Nitrogen

The deposition of nitrogen can [5]

- Injure trees by direct gas damage or by photochemical oxidants, such as ozone
- Cause acidification of forest ground so that dissolution can take place
- Contribute to water and water systems acidification
- Cause eutrophication in coastal areas and in open inland waters
- Be harmful to human health

1.6.3. Effects of Air Impurities on Human Health

Airborne particles can affect human health in many ways [2]. Certain pollutants may be toxic or carcinogenic (such as pesticides, lead or arsenic). Particles may absorb certain chemicals, and intensify their effects by holding them in the lungs for longer periods of time. Particles in the range $0.1–3\,\mu m$ can penetrate deep into the lungs.

Sulphur dioxide is soluble and is readily absorbed in the upper respiratory tract [2]. At concentrations above 1 ppm, some bronchoconstriction occurs; above 10 ppm, eye, nose and throat irritation is observed. Sulphur dioxide effects are intensified by the presence of other pollutants, especially particulates. Particulate sulphates or inert particles with adsorbed SO_2 can penetrate deep into lungs and induce severe effects.

The effects of *nitrogen dioxide* on people include nose and eye irritation, pulmonary edema (swelling), bronchitis, and pneumonia [2]. Usually, concentrations in the 10–30 ppm range are necessary before irritation or pulmonary discomfort is experienced. The oxides of nitrogen react with certain *VOCs* (reactive hydrocarbons) in the presence of sunlight to form *photochemical oxidants* [2]. Ozone (O_3) and peroxyacetyl nitrate (PAN) are two of the principal oxidising agents formed as a result of atmospheric reactions involving NO_x and reactive VOCs. These oxidants are severe eye, nose and throat irritants.

Carbon monoxide is a colourless, odourless and tasteless gas that reacts with the haemoglobin in blood to prevent oxygen transfer [2]. This is due to the higher equilibrium coefficient for the adsorption of CO onto haemoglobin compared with the coefficient for oxygen adsorption. Depending on the concentration of CO and the time of exposure, effects on humans range from slight headaches to nausea, and ultimately to death.

1.7. Preconditions for Emission Measurements

1.7.1. Need for Emission Measurements

Emission measurements are needed for many purposes. They can be used as the basis for emission studies, for dispersion model calculations and for the control of emission reduction technologies. They can also be used for the research and investigation as well as for air pollution control purposes. Legislation and control policies have been developed to improve air quality and to stimulate research efforts towards cleaner technologies. These activities rely on emission measurements.

The practical emission measurement must be performed according to the purpose of the measurement. To study the emission rates and total emissions it is essential to follow the concentrations of the most important emission gases over a long enough time period. Gas components to be followed must be selected based on the process and on the impact and rate of different emission gases. Developing emission measurement technologies make it possible to measure the concentrations of many gas components practically continuously.

For the analysis of total emissions, the volumetric flow rate of the emission gas must also be measured.

When analysing a process, there are generally many gas components to be measured. The effect of process conditions should be studied by operating the process under different process parameters, such as different capacity values. These types of emission studies produce basic results that can be applied to the control of similar processes.

1.7.2. Utilisation of Results

Results from emission measurements can be utilised for

- Air quality studies and control
- Basic data for dispersion calculations and studies
- Testing of warranty values
- Studies of the amount and quality of emissions
- Control of the performance of regulations and emission standards
- Control of the operation of gas cleaning equipment
- Control of processes

The control of processes, for instance, is essential, as it is the basic method for keeping emissions to the minimum. The development of process technologies so as to minimise emissions is the way towards permanent sustainable operation of technology-based societies.

When studying the quantity and quality of the emissions, different process conditions that can occur should be examined. It would then be possible to work out emission estimates for long periods of time and for large areas. This in turn would enable a significant reduction of emissions.

1.7.3. Importance of Process Conditions

The results obtained from emission measurements must be related to process conditions. The process to be measured should therefore be investigated in advance. The measurement plan should then be worked out based on the process data. Different processes set different preconditions to the measurement procedure. This is apparent from an example shown in Fig. 1.1.

A measurement of a short duration cannot give a reliable impression of the emission level, if the emission rate varies over time. To control the emissions, the use of a continuous measuring principle is then essential, and the duration of the measurement must be sufficient to cover all the process situations necessary for the overall emission assessment. Otherwise, the frequency and number of the measurements or the sampling selected must be high enough to ensure sufficient basis for the emission analysis.

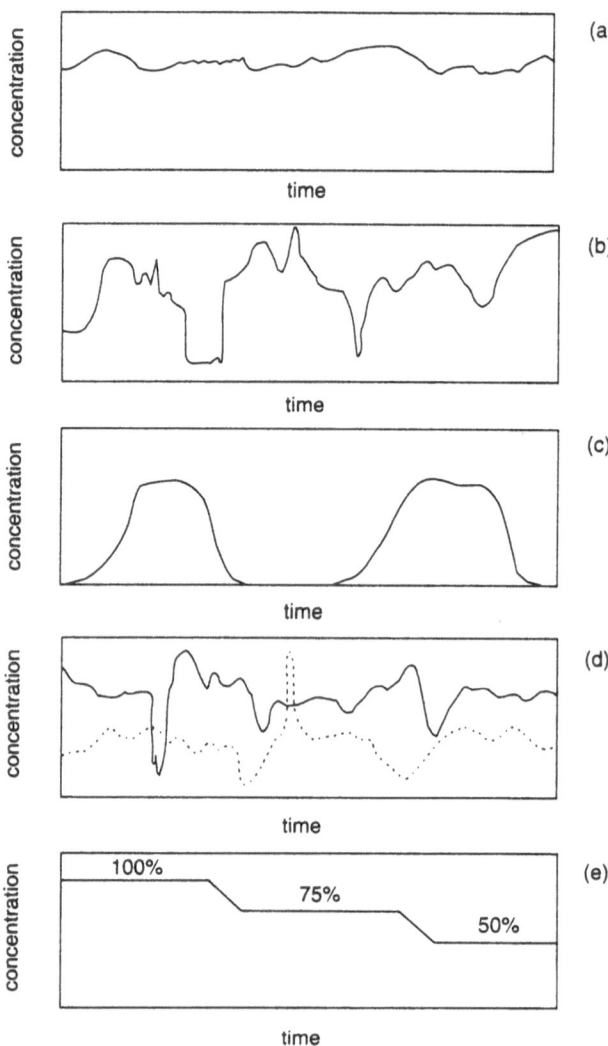

Figure 1.1. Different temporal variations of emissions. They impose different requirements on the performance of the measurements. **a** The concentration of the emission gas varies only slightly. **b** The concentration of the emission gas varies strongly. **c** The concentration of the emission gas varies periodically and noncontinuously. **d** The emission gas contains several emission components at various concentrations. **e** The concentration of the emission gas is dependent on the capacity of the process.

As the emission measurement instruments measure the concentration of an emission gas component, it is necessary to measure also the volumetric flow of the emission gas. This can be a difficult task if the outlet ducts are complicated. Based on the flow rate of the emission gas and the concentration of a certain gas component, it is possible to determine the total emission.

1.7.4. Gas Velocity Measurement

The velocity of the flowing gas can be determined based on dynamic pressure. The dynamic pressure from humid, hot gas streams containing particulate material can be measured using pitot tubes or anemometers like thermal mass flow meters. The gas velocity can also be measured based on the ultrasound velocity as measured using transmitter–receiver pairs. The static pressure inside the duct must also be determined using a "U" tube manometer or some other appropriate instrument.

A standard pitot tube can be used to obtain gas velocity profiles in the range of about 2–10 m s^{-1}. A type S pitot tube is easy to construct. Pitot tubes must be calibrated.

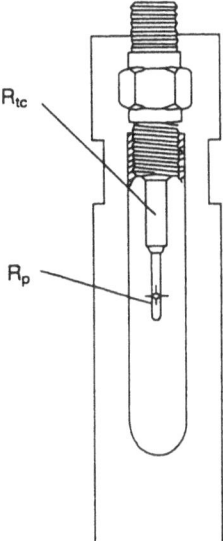

Figure 1.2. Thermal mass flow sensor [27].

The construction of a thermal mass flow sensor is presented in Fig. 1.2. The sensor consists of a platinum mass velocity sensor winding (R_p) and a platinum ambient temperature sensor winding (R_{tc}) [27]. Both are wound on a ceramic mandrel and given a protective glass coating. The sensor assembly is inserted into a stainless steel sheath. The velocity winding near the tip of the sensor is heated to a temperature above the temperature of ambient gas stream, so that the difference between these temperatures remains constant. The sensor R_p responds to mass flow by sensing the cooling effect of gas molecules as they pass the sensor. The ambient temperature sensor R_{tc} is an integral part of the air velocity sensing circuit, and compensates the circuit over a wide range of gas temperatures. The sensors provide a signal inherently corrected to standard conditions because of the temperature compensation and the calibration [27].

1.7.5. Basic Methods of Emission Measurements

Basically, emission measurements can be done using either of the two different sampling methods:

- The extractive (or diluting) method, by which the sample gas from the exhaust channel is led to the analyser (or to be taken to the laboratory) through a sampling tube line.
- The *in-situ* method, whereby a sensor is placed in the outlet pipe or the measurement is performed across the pipe, as can be done when applying optical methods.

Measurement principles are illustrated in Fig. 1.3. An example of parameters to be determined for an air emission analysis of a power plant is given in Fig. 1.4.

1.7.6. Preparative Measures

The purpose of the emission measurement to be performed, the emission gas components to be measured, and the appropriate measuring methods should be considered and selected in advance. The location and accessibility of the measuring site, the conditions of the process to be measured, fuels used, connections available and required, and other important things should be considered. The measuring site and conditions should be inspected prior to the measurement, if possible.

Preparations for the emission measurement should be completed in the laboratory before moving to the measurement site. Instruments should be calibrated, possible filters weighted, and the necessary measuring instruments and accessories properly packed for the trip. It is useful to work out a list of devices, and use it to check that all the necessary instruments and accessories are available.

Note should be made of the process data relevant to the measuring time period. It is often helpful to make use of process data forms used in process plants or emission data forms used by the officials. There are instructions available for the preparation for the emission measurements. Table 1.3 lists a test schedule and personnel responsibilities with an example, published by the American EPA (Environmental Protection Agency) [29].

1.7.7. Elimination of Disturbances Caused by Water Vapour

When using continuous measurement instruments for the measurement of emission gas concentrations, disturbing factors should be eliminated. For example, to remove particulate material from the sample gas flow, filters are generally used.

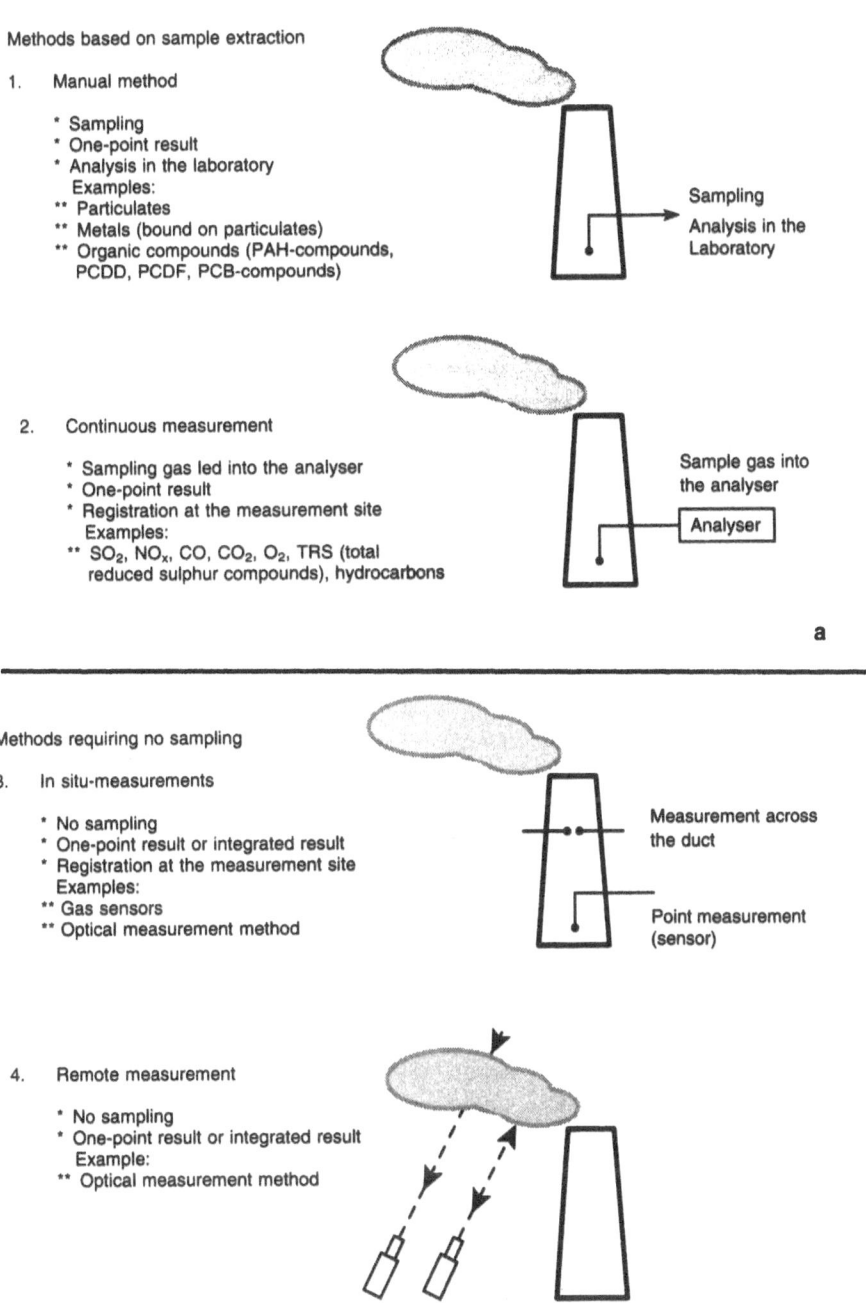

Methods based on sample extraction

1. Manual method

   * Sampling
   * One-point result
   * Analysis in the laboratory
     Examples:
   ** Particulates
   ** Metals (bound on particulates)
   ** Organic compounds (PAH-compounds, PCDD, PCDF, PCB-compounds)

   Sampling
   Analysis in the Laboratory

2. Continuous measurement

   * Sampling gas led into the analyser
   * One-point result
   * Registration at the measurement site
     Examples:
   ** SO_2, NO_x, CO, CO_2, O_2, TRS (total reduced sulphur compounds), hydrocarbons

   Sample gas into the analyser

   Analyser

   a

Methods requiring no sampling

3. In situ-measurements

   * No sampling
   * One-point result or integrated result
   * Registration at the measurement site
     Examples:
   ** Gas sensors
   ** Optical measurement method

   Measurement across the duct

   Point measurement (sensor)

4. Remote measurement

   * No sampling
   * One-point result or integrated result
     Example:
   ** Optical measurement method

   b

Figure 1.3. Basic methods of emission measurements.

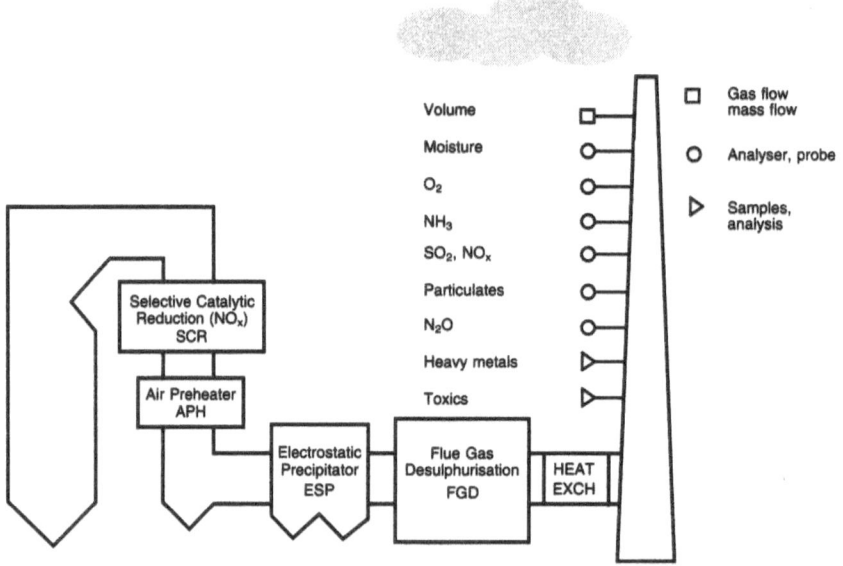

Figure 1.4. Parameters of an air emission analysis of a power plant [28].

The possible interference caused by water vapour normally present must also be eliminated. Dilution of the gas sample using dry air is often an appropriate solution. If water is removed from the sample gas, there is always a chance that part of the gas component to be measured is thereby also removed.

Sometimes it is, however, necessary to remove water from the sample gas stream, or the sample gas must be dried. Electrical dryers, for instance, cool the sample gas down to 2–5 °C. Water vapour condenses at this temperature, and would not do so in the gas pipes following, as these are maintained at higher temperatures. Not all the water is removed from the sample gas using this technique, but the moisture content of the gas passed through the dryer is relatively constant and can be taken into account. One way of realising the cooling is the use of Peltier elements.

Other methods applicable to the drying of the sample gas include permeation drying and cooling by water. Permeation dryers use polymeric membranes, through which water is selectively diffused. The outer surfaces of the permeation tubes are continuously dried using dry gas flow, so that the gas flowing inside the tubes is dried, and the gas at the outer surface flowing in the opposite direction is humidified [30].

In measurements of a short duration, an ice–water bath can be used for the cooling. This is an isolated container having a dryer-coil in the bath. One can also use just a water bath or a dryer circulating cooling water. In short measurements, the water contained in the sample gas can also be removed

Table 1.3. Test preparations [29]

Describe and identify the following:

- Construction of special sampling and analytical equipment

  Description
  Dates for completion of work
  Responsible group

- Modifications to the facility, e.g. adding ports, building scaffolding, installing instrumentation, and calibrating and maintaining existing equipment

  Description
  Dates for completion
  Responsible group

- Services provided by the facility, such as electric power, compressed air, and water

  List of all services to be provided by the facility
  Description of modifications or added requirements, if necessary

- Access to sampling sites

  Description
  If modifications are required, requirements and responsible group

- Sample recovery area

  Description
  If a mobile recovery area or laboratory is used, installation location, dates for installation, and responsible group

---

Example

Test preparation

1. *Construction of special sampling and analytical equipment.* There are no equipment modifications or special analytical equipment required for this site.

2. *Modifications to facility.* The [*Plant*] crew will install additional 4-inch ID sampling ports as shown in Figures enclosed. In addition, the decking at the outlet stack will be extended to circumvent the stack to allow access to the new sampling port locations. All work will be completed during the scheduled plant shutdowns on July 11 and 25, 1991.

3. *Services provided by the facility.* The [*Plant*] agreed to furnish additional temporary 110 volts, 20 amp power as follows:

   - EFB (Electrified filter bed) inlet   5 outlets
   - EFB outlet stack   5 outlets
   - Press vents   2 outlets
   - Mobile CEM (continuous monitoring) lab   5 outlets

4. *Access to sampling sites.* There are no special problems or safety issues in gaining access to the testing locations.

5. *Sample recovery areas.* [*Contractor*] will provide an office trailer (32 ft, 2 foot tongue) and a smaller trailer for sample recovery areas. The office trailer requires a single phase 220 volt power supply for lighting and air conditioning and the smaller trailer requires two 110 volt, 20 amp circuits. The sample recovery task leader will be responsible for locating both sample recovery units in areas as free as possible from ambient dust contamination. The office unit will be used for recovering the M202 and MM5 samples, and the smaller unit will be used for the M0011 (formaldehyde) samples.

---

using different drying agents [30]. The most common of these are silica gel, different molecular sieves and phosphorus pentoxide. Drying agents in gas sampling lines are normally packed in bubbling containers, drying towers, or drying cartridges.

The most frequently used drying agent is silica gel, which is available in different grain sizes. It can be used for the drying of most gases (except HF) [30]. This material can also be quickly recovered. In 10 min at 200 °C, 80% of the moisture bound by silica gel is removed.

The condensation of water vapour in the gas sampling can be prevented in two ways. The sample gas can be diluted using dry air, which enables the realisation of the analysis at room temperature without the moisture being condensed. The second option is to pass the sample gas to the analyser without dilution, and to keep the sample line and the analyser at the higher temperature of the sample gas.

1.7.8. Calibration

The calibration of a continuous emission measurement instrument system is necessary to find the real correlation between the instrument display and the concentration of the emission gas component to be measured [31]. The linearity of the instrument is investigated by using a multipoint calibration. To determine the calibration curve, four different calibration gas concentrations in addition to zero-gas are often considered sufficient. The concentrations should evenly represent the whole concentration range of the actual measurement.

The determination of the calibration curve is quite a time-consuming task. It must be done, however, at certain time intervals, as the characteristics as well as the performance of the measuring instruments must be known and followed. In practical field measurements, the frequent calibration is often performed by checking the display values for only two concentrations (two-point calibration), one for the zero-gas (e.g. nitrogen gas) and the other for an appropriate concentration of the calibration gas. The linearity characteristics must then be known from earlier calibrations. Included in the calibration procedure, the tightness, response time, stability and sensitivity are also checked [32].

When using reactive gas components, like NO and NO_2, some problems can arise from the possible instability of the gas composition used as the calibration gas. Estimates of the stability periods of certain calibration gas compositions are presented in Table 1.4. In the same table, minimum values of the operation pressures, as defined by the suppliers of the calibration gases, are presented. These limits are based on the fact that there could be changes in the concentration of the calibration gas if the total pressure in the gas cylinder becomes too low.

Table 1.4. Stability periods and minimum useful pressures of some calibration gases [33]. Concentration values are expressed as volume parts per million (vpm)

Gas component	Concentration range (vpm)	Stability period (months)	Minimum cylinder pressure (bar)
H_2S	10–100	6	10
	>100	12	5
NO	0.5–2000	12	5
NO_2	1–100	12	10
	100–5000	12	5
H_2O	1–10	3	50
	10–100	6	50
SO_2	0.1–100	12	10
	100–5000	12	5
NH_3	10–2000	12	5
CO	1–1000	12	5
	>1000	12	2

Chapter 2

Determination of Mass Concentration of Particulate Material

2.1. Methods Based on Sampling: Gravimetric Method

The gravimetric method is based on the sampling of the flowing particulate-loaded gas from different points across the exhaust gas duct, and the gravimetric determination of the mass of the particulate material. The sample is collected over a certain period of time from each point. The volumetric gas flow is measured. To get the result, the following steps must be performed [34]:

- Determine the volumetric flow of the gas.
- Take the particulate material sample (on filter).
- Adjust the sample gas flow (isokinetic sampling).
- Determine the volume of the sample gas.
- Weigh the particulate material.
- Calculate the concentration and the mass flow.

The method is applicable to the determination of the concentration (and emission) of the particulate material from a mixture of gas and particulate material flowing through a known cross-sectional area of a flue or emission pipe or duct.

A general procedure for the determination of the concentration of particulate material is presented in Fig. 2.1.

2.1.1. Gas Sampling

Sampling for the determination of the mass concentration of particulate material is schematically presented in Fig. 2.2. Sampling units have been developed towards better practicability. Figure 2.3 shows the construction features of a sampling unit for the determination of the concentration of particulate material.

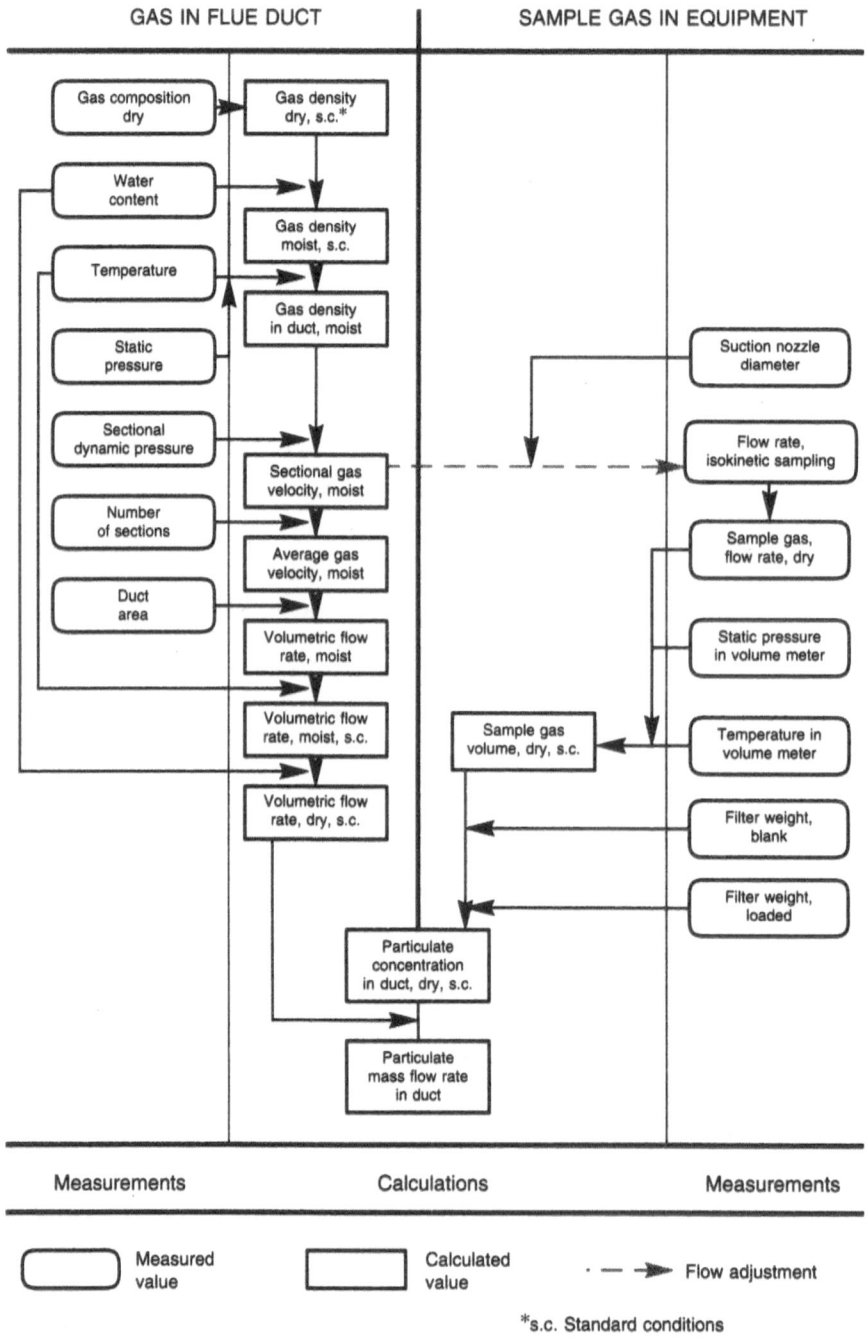

Figure 2.1. A general principle of the determination of the concentration of particulate material [34, 35].

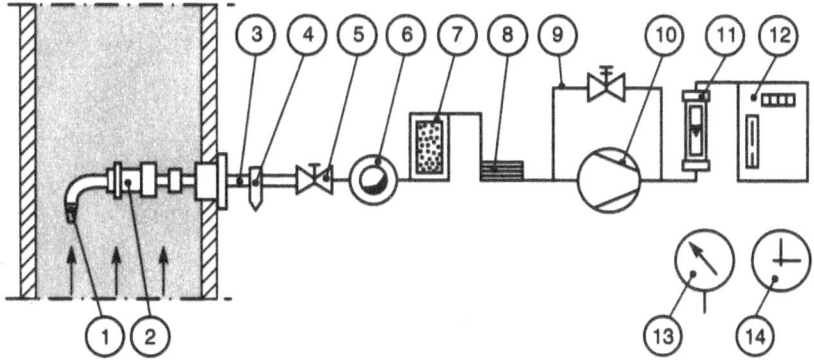

Figure 2.2. Particulate material sampling ("in-stack" sampling, VDI 2066). 1, sampling probe; 2, in-stack filter holder; 3, sampling tube; 4, direction indicator; 5, closing valve; 6, condensed water separator; 7, dryer tower; 8, pump seal; 9, by-pass and valve; 10, pump; 11, rotameter; 12, gas meter and thermometer; 13, air pressure gauge; 14, clock.

2.1.2. Isokinetic Sampling

In order to get a representative sample for the determination of the concentration of particulates, it is essential to be able to regulate the sample gas stream according to the flow rate of the main gas stream. This is done by isokinetic sampling. Deviation from the isokinetic sampling may result in the sample not being correctly representative. This is illustrated in Fig. 2.4.

In practice, it is not possible to obtain an exact particulate sample distribution even by isokinetic sampling [37]. This is because of effects such as gravitational sedimentation, surface drag, presence of an impaction surface (even though probe ends are tapered, sharp and smooth), particle coalescence and evaporation, gradients of concentration, temperature, pressure and other effects.

A 10% discrepancy in gas velocity between the flow at the nozzle tip and the flow in the duct can cause a deviation of greater than 10% in the value of the concentration measurement. This error, however, strongly depends on particle size and gas velocity [35]. If the gas is sampled isokinetically and the angle between the flow direction and the nozzle axis is not more than 15°, the deviation from an aligned nozzle will be less than 3.5% [35].

2.1.3. Number of Sampling Points

The number of sampling points should be selected based on the size of the duct to be sampled and the distance between the sources of flow disturbance and the

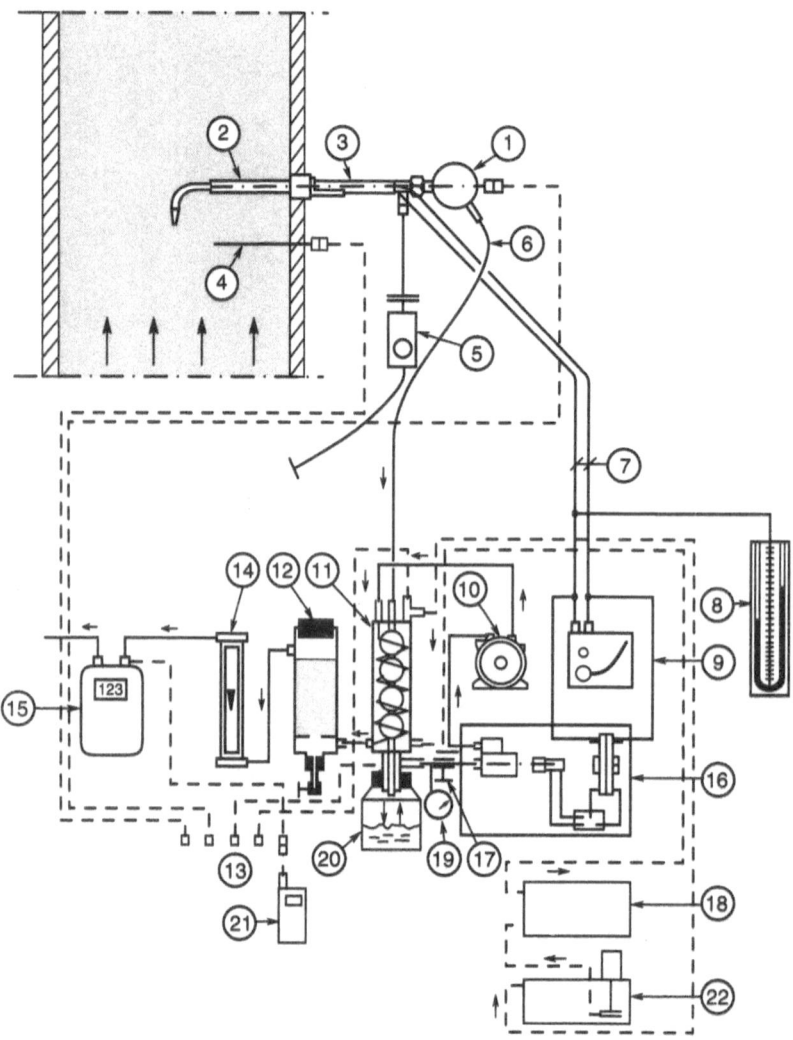

Figure 2.3. Constructional scheme of a particulate material sampling unit [36]. Parts of sampling unit: 1, plane filter; 2, sampling probe; 3, probe cover; 4, temperature sensor; 5, filter temperature controller; 6, vacuum tube; 7, tubing for pressures of probe and duct. Parts of measurement unit: 8, pressure gauge for duct; 9, zero pressure indicator; 10, vacuum pump; 11, glass double cooler; 12, dryer tube; 13, operation panel; 14, sample gas flow meter; 15, flow counter; 16, motor valve for automatic isokinetic sampling; 17, manual control valve; 18, compressor cooler; 19, vacuum gauge; 20, condensed water container; 21, digital thermometer; 22, cooling water container and circulator pump.

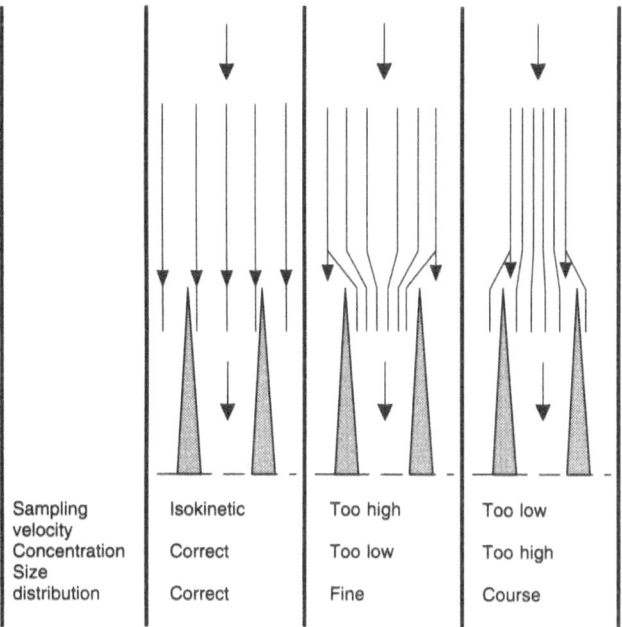

Sampling velocity	Isokinetic	Too high	Too low
Concentration	Correct	Too low	Too high
Size distribution	Correct	Fine	Course

Figure 2.4. Effect of the sampling velocity on the sample distribution by the particle size [34].

sampling level. In Fig. 2.5, an example is given, based on a national standard, of the number of sampling points dependent on the size of the duct, the distance from a flow disturbance before the sampling level, and the undisturbed duct length after the sampling level. In the figure,

- M_1 is the undisturbed distance before the sampling level.
- M_2 is the undisturbed distance after the sampling level.
- D is the (hydraulic) diameter of the channel.

For example, if the diameter of a circular duct is $D = 1000$ mm, $M_1 \geq 5D$ and $M_2 \geq D$, the number of sampling points based on Fig. 2.5 is 11. We select 12, as the number should be even. The sampling should be performed according to valid standards.

In the example given in Fig. 2.5, the location of the sampling points is determined on an equal area basis. The traverse points at the centroids of the equal areas of the circular stack indicated are used as the sampling points. Some standards include the central point of the duct in the sampling points.

The hydraulic diameter is sometimes used, as the ducting is often rectangular or square rather than circular because of space limitations, for ease of installation, and so forth [2]. The hydraulic diameter (D_h) can be calculated as $D_h = 4A/P$, where A is the cross-sectional area and P is the inside perimeter of the duct.

In Nordic countries, use is generally made of an isokinetic sampling based on a "zero-indication", whereby the kinetic and static pressures are measured

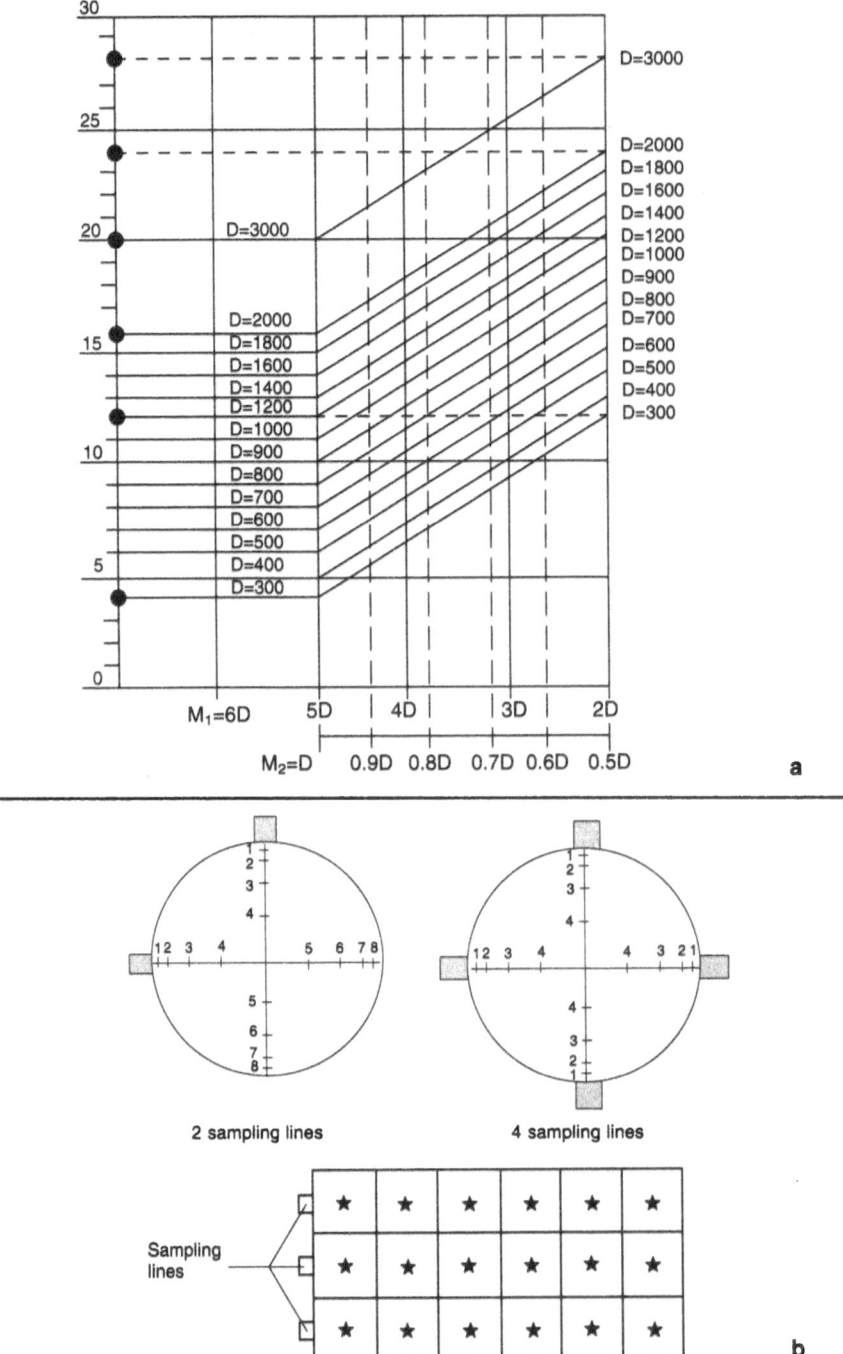

Figure 2.5.a Dependence of the total number of the sampling points on the location of the measurement level and the size of the channel. **b** Two examples of sampling locations based on equal areas [34].

connected with the sampling probe. The variation in the difference between these two pressures indicates the variation in the gas flow rate, and is used for the regulation of the isokinetic sampling. In the USA and Canada, for instance, another system for the control of the isokinetic sampling is used, where an "S" type pitot tube by the side of the sampling probe measures the velocity of the gas at the measuring point. Based on the measurement data, the isokinetic sample flow is calculated. The method is in accordance with the EPA standards.

2.1.4. Selection of Suitable Sampling Location

The sampling location should be situated in a length of straight duct with constant shape and constant cross-sectional area, preferably vertical, and downstream as far as practieable from any obstruction which may cause a disturbance and produce a change in the direction of flow (e.g. a bend, a fan or a partially closed damper) [35].

To ensure a sufficiently homogeneous gas velocity distribution in the sampling plane, this section of straight duct should be at least seven hydraulic diameters long. Over the length of the straight section, locate the sampling plane at a distance of five hydraulic diameters from the inlet. If the sampling plane is to be located in a chimney discharging to the open air, the distance to the chimney top should also be five hydraulic diameters [35]. According to some standards, the undisturbed distance for the flow before the sampling location must be at least five hydraulic diameters, and after the measurement level at least one hydraulic diameter. If this condition cannot be met, the ratio of the undisturbed distances before and after the sampling location should be selected to be 5:1 [34].

2.2. Automatic Particulate Monitoring

For the continuous monitoring of the particulate concentration in emissions, continuously operating instruments have also been constructed [38]. Some constructions sample flue gas isokinetically with the retention of particulates on an automatically advanced sequential filter tape. The mass of the particulate deposit on the filter tape is measured differentially before and after sampling by a gauge based on the absorption of beta radiation by matter.

Another construction adapts the tapered element oscillating microbalance method to determine particle emissions based on mass concentration in real time [39]. The sample is extracted from the process through a sampling cyclone to define upper size cut for particles (Fig. 2.6). The sample is diluted and cooled with a two-stage dilution system utilising ejector-based dilution units.

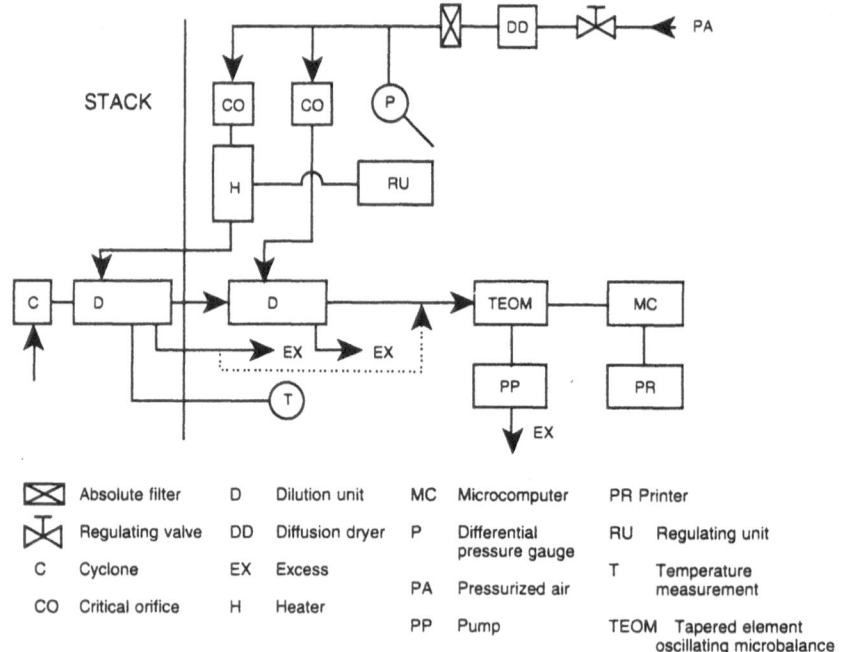

Figure 2.6. Scheme of the particle emission monitor based on the tapered element oscillating microbalance [39].

The system makes it possible to measure aerosol particulate mass concentration in real time [39]. The combustion aerosol must be diluted using clean, dry air with a known dilution ratio. The microbalance uses a filter on a tapered tube to collect particles from the gas flow through the tube. The tube is maintained in oscillation by the feedback electronics. The oscillating tube with the filter on its upper end can be considered as a simple harmonic oscillator. The frequency of the tube changes when particles are collected on the filter. A precision frequency counter measures the frequency at close intervals and the results are fed to a microprocessor which, with the value of the flow through the tube, can evaluate the mass concentration.

Automatic monitoring of mass concentrations of particles must be correlated with the manual method.

2.3. Optical Measurement Method

It is often desirable to follow the emissions of particulates continuously. This is important for the control of processes, so as to improve the process performance and to keep particulate emissions low. For rapid on-line control of particulate material, optical methods are often used. These methods are generally based on the attenuation of light passing through a particle population. In addition to continuous operation, one of the advantages of optical methods is that they do not interfere with the process to be measured.

The attenuation of light by particles is caused by the interactions between particles and light, i.e. the absorption and scattering of light. The interpretation of these phenomena, and the transformation of the information included in them into a signal indicating particulate concentration, is complicated and dependent on many factors. The control of particulate emissions using optical methods can be best performed when the quality of dust and possibly the particle size distribution of particulate emissions remain relatively constant.

2.3.1. Basis of Measurement

We can start the analysis of the interaction between light and a particle population by considering the interaction between one particle and light, which is a form of electromagnetic radiation. This interaction takes place in a medium, which is emission gas in this case. There are several mechanisms in the interaction, appearing basically in two ways as observed in the behaviour of light [40]:

1. *Absorption*. This is the attenuation of the radiation intensity inside the particle, through the transfer of energy from the radiation into the material, and is associated with an excitation into a state of higher energy inside the material. Absorption can occur when there is a wavelength in the incident light compatible with an absorption band of the particulate material.
2. *Scattering*. This is where radiation is emitted from the surface of a particle in different directions.

As is evident from the above, both the absorption and the scattering of light are due to the interaction between electromagnetic radiation and matter. The absorption can be loss of radiation energy as consumed during the excitation of an electron in an atom, as is often in the case in the absorption of ultraviolet or visible light. The absorption can also be caused by the change of the energy associated with the vibrations between atoms in a molecule, caused by a light beam. It can also be due to the change of the rotational energy of a molecule by the influence of light. Vibrational and rotational energy changes can be associated with the absorption of infrared light, the energy of which is not sufficient for the excitation of electronic states.

In absorption, light gives up energy and is consequently attenuated. In Fig. 2.7, the absorption is schematically presented by showing a photon absorbed, which excites an atom to a higher state of energy. In a semiconducting solid material, the absorption can cause some of the outer shell electrons of an atom (valence electrons) to escape from the atom and be transferred into the so-called conduction band, which is a higher energy state

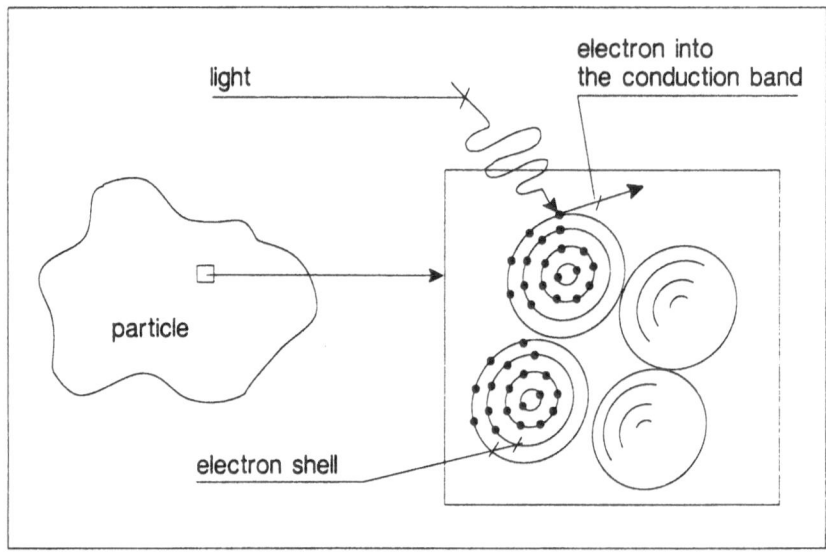

Figure 2.7. An atom in an electromagnetic field can absorb radiation and be simultaneously excited.

compared with the basic energy state. The energy absorbed may also be sufficient to ionise an atom by giving kinetic energy to an electron. This is photoelectric emission.

Scattering presupposes the interaction of electromagnetic radiation with an electrical quantity. This electrically charged "particle" is naturally a cloud of electrons bound to an atom. The electromagnetic radiation, in this case light, makes the electron cloud vibrate, and this vibrating electric charge generates its own radiation, which is the scattered light referred to. The generation of the phenomenon of scattering is represented schematically in Fig. 2.8. The example used is a sufficiently simple model of a hydrogen atom. In this case it is assumed that the electron cloud around the nucleus is a sphere, inside which the charge is uniformly distributed.

Figure 2.8 shows a situation where the centre of charge in the electron sheath has instantaneously been transferred from the nucleus to distance r by the influence of electromagnetic radiation (light). The influence of light makes the electron sheath vibrate in relation to the nucleus. This electric oscillator generates an electromagnetic wave motion or the scattering light.

The characteristic frequency of the electron in a hydrogen atom can be estimated on the basis of the Bohr atomic model. In that model, the electron is thought to circulate around the nucleus in its own orbit, whose radius is R_B. The electron remains in its orbit due to the equilibrium between the attractive force of the nucleus and the centrifugal force. The electric attraction of the nucleus in the case of the hydrogen atom is $e^2/(4\pi\varepsilon_0 R_B^2)$. By equating it with the centrifugal force we obtain

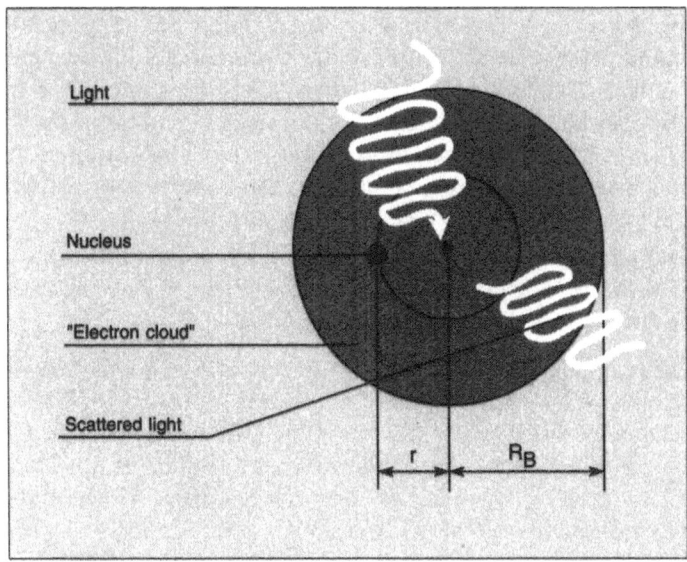

Figure 2.8. Model of a hydrogen atom and its contribution to scattering of light.

$$m_e \omega^2 R_B = \frac{e^2}{(4\pi\varepsilon_0 R_B{}^2)} \tag{2.1}$$

Here ε_0 is the permittivity of a vacuum, $8.8854 \times 10^{-12}\,\mathrm{F\,m^{-1}}$, m_e is the electron mass, $9.1 \times 10^{-31}\,\mathrm{kg}$ and ω is the angular velocity.

By substituting into (2.1) the approximate radius of the hydrogen atom, that is 0.05 nm, we obtain a value for the angular velocity ω roughly $4.5 \times 10^{16}\,\mathrm{s^{-1}}$, the vibration frequency being about $10^{16}\,\mathrm{Hz}$. This frequency is that of ultraviolet light. The frequencies of visible light range roughly from $4.2 \times 10^{14}\,\mathrm{Hz}$ for red light to $7.5 \times 10^{14}\,\mathrm{Hz}$ for blue light, that is to say, they are lower than a tenth of that for a hydrogen atom. A calculation shows that the energy of visible light is not sufficient to excite this atom. In other words, the light is not absorbed, but scattered. As is apparent from above, for example, the electrons in the outer shells of atoms of solid particles, or the valence electrons, can be detached by the action of light, when absorption of light takes place. Thus, in the case of solid particles both absorption and scattering of light can occur.

Scattering is an incoherent phenomenon and it is generally not associated with interference phenomena, which amplify light in certain directions. Through scattering, part of the intensity of the light is therefore dissipated by the action of particles, and this can be utilised in the measurement of particle concentration.

The scattering of electromagnetic radiation can be divided into two types according to the interrelation between the wavelength of the radiation and the particle size [40]:

1. Scattering is Rayleigh scattering when the diameter of the particle is small compared to the wavelength of the radiation (in this case the diameter of the particle is less than $0.05\,\mu m$). The intensity of this scattering is inversely proportional to the fourth power of the wavelength, so that in the scattered radiation short wavelengths are dominant [41]. The blue colour of the sky is due to this phenomenon, since the ratio of the intensities of scattered blue and red light, for instance, is $(4.2/7.5)^{-4}$, or approximately 10.

2. Mie scattering takes place when the particle diameter is comparable with the wavelength of the incident radiation. The diameters of particles in emission gases are usually in the range of Mie scattering.

The intensity of Mie scattering depends on the ratio of the particle diameter to the radiation wavelength, as well as on the optical properties of the particulate material, or more specifically, on the ratio of the complex refractive index of the particle to that of the medium. The medium is not assumed to absorb light, so its refractive index has only the real part. Knowing these two ratios is sufficient for the use of the Mie theory [42]. The ratio of the particle diameter to the radiation wavelength is defined by the so-called diffraction parameter x [40]:

$$x = \frac{2\pi a}{\lambda} \tag{2.2}$$

where a is the geometrical radius of the particle and λ is the wavelength of the radiation.

The optical properties of the particulate material, on the other hand, are defined by the complex refractive index m, that is,

$$m = n - ik \tag{2.3}$$

where n is the refractive index of the particulate material, k is the attenuation coefficient of the particulate material and i is the imaginary unit ($\sqrt{-1}$).

k is dependent on the wavelength of radiation for two reasons, since

$$k = \frac{\alpha\lambda}{4\pi} \tag{2.4}$$

where $\alpha = \alpha(\lambda)$ is the linear absorption coefficient of the particle.

Figure 2.9 shows the optical constants n and k of sodium sulphate as a function of the wavelength or the wavenumber.

The Mie theory gives the absorption and scattering efficiency of a particle. It is these quantities that determine the attenuation or extinction efficiency (Q_{ext}), or

Attenuation = Absorption + Scattering

The calculation of attenuation, absorption and scattering efficiencies is very laborious, as they are series developments of the complex Bessel and Hankel functions. Computer technology has made it possible to calculate these efficiencies accurately only since the 1980s.

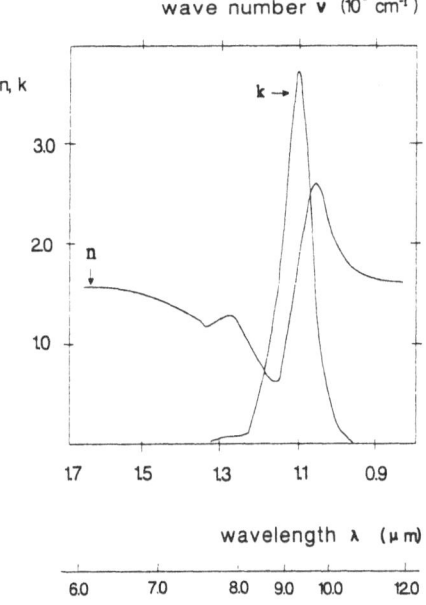

Figure 2.9. Optical constants *n* and *k* of sodium sulphate. In the region of visible light the value of the refractive index is 1.48 [40].

The Mie theory also yields the scattering pattern of the particle, or the relative intensity of the light dependent on the scattering angle. The scattering angle is the angle between incident and scattering radiation. Fig. 2.10 shows the distribution of Mie scattering, when the Mie parameters of the particle are $x = 6.0$, $n = 1.84$ and $k = 0$.

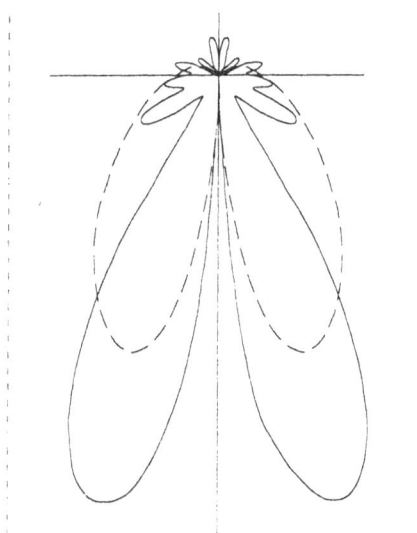

Figure 2.10. Distribution of the Mie scattering of a particle, when $x = 6.0$, $n = 1.84$ and $k = 0$. The distributions have been weighted with the sine function of the scattering angle [42]. The incident radiation is directed downwards. The dashed line represents another theory, the distribution in accordance with what is called the Henyey–Greenstein theory.

The Mie theory has restrictions when used alone. It does not allow for the influence of other surrounding particles, which may cause re-scattering of the scattered radiation. Furthermore, it does not allow for the interference of the radiations scattering from various particles. This shortcoming is perhaps more serious than the former. For the Mie theory to be accurate, the distance between the particles should be at least three times the diameter of the particle.

An example of the dependency of the attenuation efficiency of light radiation on the diffraction parameter is shown in Fig. 2.11.

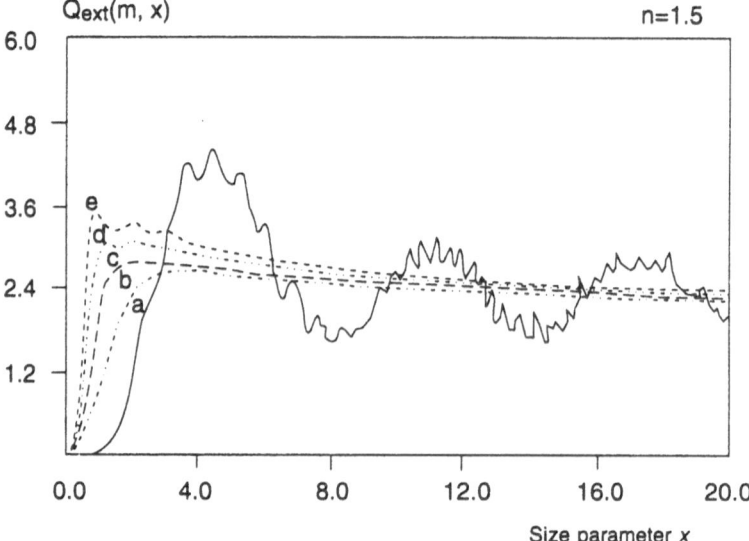

Figure 2.11. Extinction efficiency Q_{ext} as a function of the diffraction parameter x. Curve **a** represents a situation where the particulate material is sodium sulphate and measurement wavelength is in the visible region of light. Curves **b**, **c**, **d** and **e** represent approximately the attenuation efficiencies of the ash of various coal qualities when using visible and near infrared light [43].

Figures 2.12 and 2.13 show the attenuation efficiencies as a function of the wavelength for monodispersive sodium sulphate dust and the fly ash of coal. The attenuation efficiencies have been calculated for several different particle sizes. It is, therefore, to be noted that the quality of the dust determines the attenuation of light.

The concentration of the particulate material in an emission gas can be measured on the basis of both scattering and attenuation. The measurements based on scattering have been found, theoretically as well as from practical experience, to be more unreliable than measuring methods based on attenuation. Instruments based on scattering have, however, the advantage of being easy to align, because the detector can be fixed to the radiation emission unit.

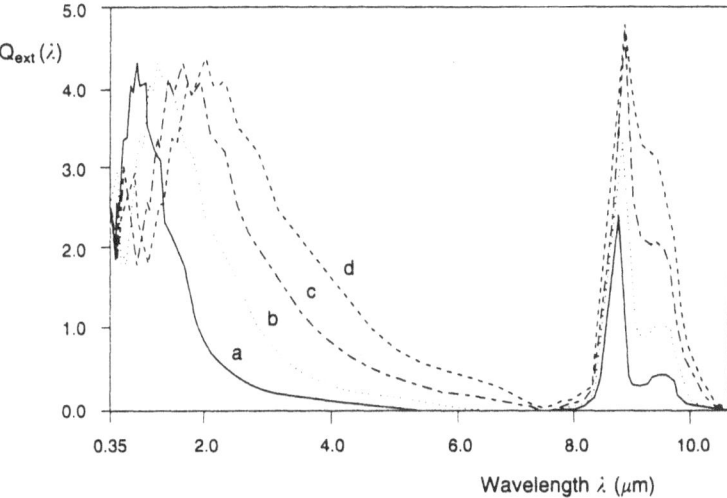

Figure 2.12. Extinction efficiency Q_{ext} as a function of the wavelength for sodium sulphate particles the radii of which are **a** $0.75\,\mu m$, **b** $1.0\,\mu m$, **c** $1.25\,\mu m$ and **d** $1.5\,\mu m$ [40].

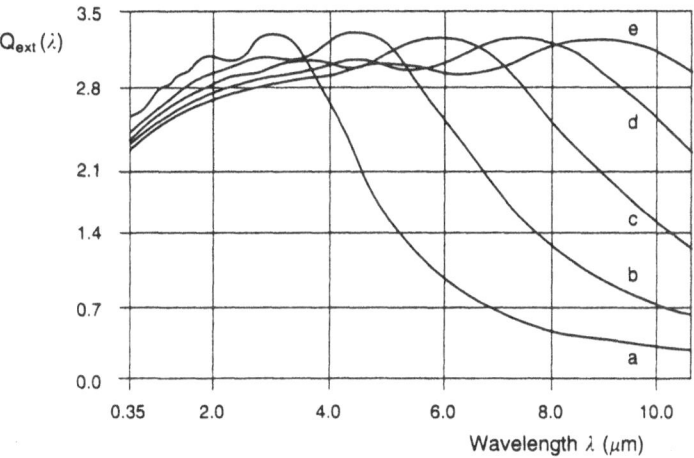

Figure 2.13. Extinction efficiency Q_{ext} as a function of the wavelength for particles of the fly ash of coal the radii of which are **a** $0.75\,\mu m$, **b** $1.0\,\mu m$, **c** $1.25\,\mu m$, **d** $1.5\,\mu m$ and **e** $1.75\,\mu m$ [40].

2.3.2. Calculation of Concentration Based on Light Attenuation

The arrangement for the measurement of the concentration of solid particles in emission gases based on light attenuation is shown simplified in Fig. 2.14.

A particle population in an emission gas duct can be represented by the particle concentration N_0 and the the distribution function $p(a)$ [43]. For instance, the sodium sulphate dust produced by a soda recovery boiler follows fairly well the log-normal distribution.

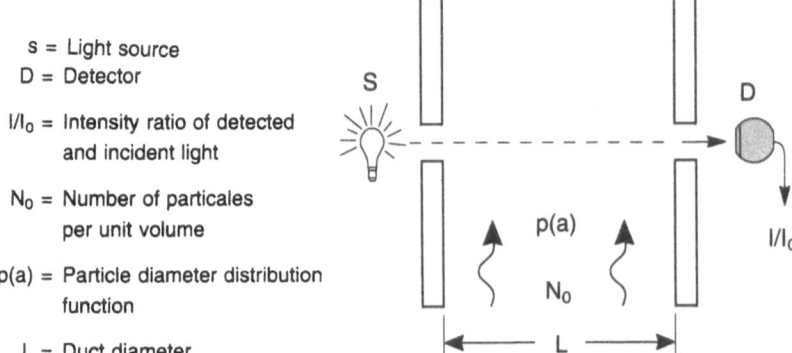

Figure 2.14. Measurement of the concentration of particulate material based on light attenuation [40].

The attenuation of light by the action of dust particles in the gas duct follows the Lambert–Beer law, a very well known law which is generally applied to the attenuation of radiation by the action of matter. According to the law

$$\frac{I}{I_0} = e^{-\gamma L} \tag{2.5}$$

where I is the intensity of the light which has penetrated the duct, I_0 is the intensity of the light incident on the duct, γ is the exponential attenuation coefficient and L is the width of the gas duct, that is, the distance light travels in the particle population.

The exponential attenuation coefficient γ can, for a particle population containing monodispersive, spherical particles, be written in the form [40]

$$\gamma = \pi a^2 Q_{\text{ext}}(m, a, \lambda) N_0 \tag{2.6}$$

For a polydispersive particle population, the examination has to be extended over all sizes of the particles, yielding

$$\gamma = \sum_i Q_{\text{ext}}(m, a_i, \lambda) \pi a_i^2 N_i \tag{2.7}$$

where N_i is the number of particles with radius a_i.

$$\sum_i N_i = N_0$$

By using the distribution function $p(a)$ of the particles it is possible to integrate over the whole particle population, and the final form of the exponential attenuation coefficient can be represented as follows:

$$\gamma(m, \lambda, p(a)) = \pi N_0 \int_0^\infty Q_{\text{ext}}(m, a, \lambda) a^2 p(a) \, \mathrm{d}a \tag{2.8}$$

It appears from the equation that the exponential attenuation coefficient is a function of at least three unknown factors. The unknown factors which would have to be solved by means of measurements are N_0 as well as the parameters determining the distribution function of the particles, such as the mean radius and the standard deviation of the distribution.

On the other hand, the concentration of particles can be represented in the form

$$M = \frac{4}{3}\pi N_0 \rho \int_0^x a^3 p(a)\, \mathrm{d}a \tag{2.9}$$

where ρ is the density of the particulate material.

By solving N_0 from Eq. (2.8) and substituting it in Eq. (2.9) we get the equation relating the concentration of particles to the attenuation measured:

$$M = \frac{4}{3}\rho \frac{\int_0^x a^3 p(a)\, \mathrm{d}a}{\int_0^x Q_{\mathrm{ext}}(m, a, \lambda) a^2 p(a)\, \mathrm{d}a} \frac{1}{L} \ln\left(\frac{I_0}{I}\right) \tag{2.10}$$

If the measurement is carried out by using, for example, an incandescent lamp as a radiation source, the integral in the denominator has to be replaced by a double integral where, in addition to the integration of particle size, integration over the wavelength band formed by the radiation source and the detector has to be performed.

The use of Eq. (2.10) presupposes that the optical properties and the particle size distribution of the particulate material are known. If these quantities are known, it is possible to calculate numerically the coefficient between the particle concentration and the light attenuation. This coefficient is the ratio of the two integrals as represented in (2.10).

2.3.3. Measurement Arrangement

Instruments for the measurement of the concentration of particulate material based on the attenuation of the light permeating the emission gas differ from each other mainly as far as the light source used is concerned. This is what determines the measurement wavelength band. The light source affects the reliability of the measurement results, size of measurement access ports, maintenance, and related factors.

Incandescent Lamp as Light Source

The incandescent lamp is the most commonly used light source; it is cheap and simple to use. The wavelength band emitted by the incandescent lamp almost follows the radiation of a black body. Some of the drawbacks of the incandescent lamp are the low radiation power, the difficulty of directing the radiation, and the strong decline in the radiation power with increased operating time. Figure 2.15 shows a typical spectrum emitted by a black body.

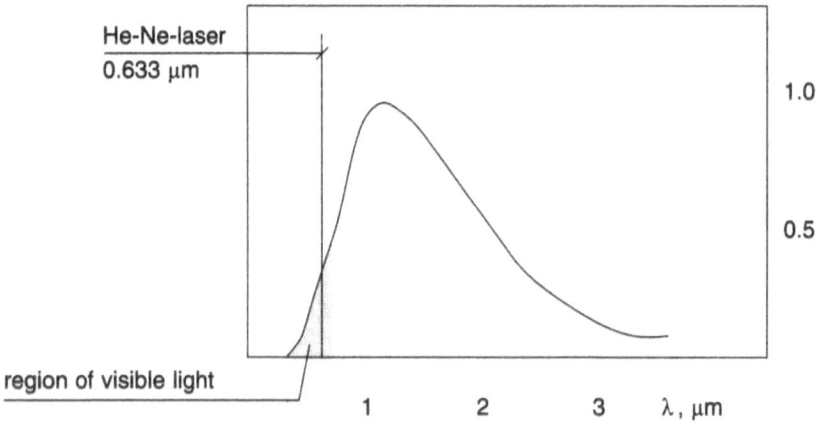

Figure 2.15. Form of the radiation spectrum emitted by a black body. This can be used to approximate representation of the wavelength band of an incandescent lamp [40].

The decrease in the radiation power of an incandescent lamp in the course of the operation time, as well as, for instance, changes caused by variations in the operating voltage can be compensated for by using the measuring arrangement shown in Fig. 2.16.

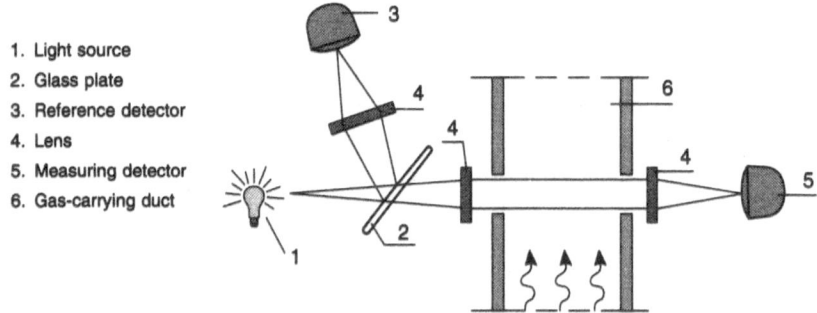

1. Light source
2. Glass plate
3. Reference detector
4. Lens
5. Measuring detector
6. Gas-carrying duct

Figure 2.16. Measuring system by means of which the variation of the radiation power of the source can be compensated [40].

The use of an incandescent lamp requires lenses so that it would be possible to focus the departing light sufficiently small for the measurement. In practice, however, the diameter of the beam is rather big, and therefore one is forced to use purge air as shown in Fig. 2.17. The use of purge air is necessary in order to get a reliable measurement result.

The measuring arrangement in Fig. 2.17 prevents the accumulation of particles in the optical components of the measuring device such as lenses, viewports and other places where they can disturb the measurement. Using purge air is a simple and common method by means of which the fouling of the optical parts is prevented. When using purge air, attention has to be paid to keeping the air flow constant, and ensuring that the temperature of the purge air is suitable. Then it is not possible for the water to condense in the optical components.

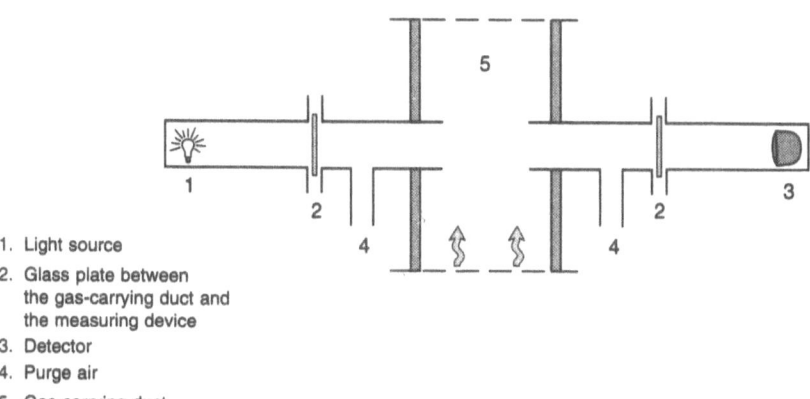

1. Light source
2. Glass plate between the gas-carrying duct and the measuring device
3. Detector
4. Purge air
5. Gas-carrying duct

Figure 2.17. It is necessary to get purge air into the measuring equipment so that the optical components would stay clean [40].

Laser as Light Source

The light source used in the laser-based equipment for measuring particle concentrations is most commonly helium–neon laser, its wavelength of radiation being $0.633\ \mu m$. This light can be assumed to be fully monochromatic.

Due to the properties of the laser [40]

1. The theoretical calculations become considerably easier, because the laser light is monochromatic (cf. incandescent lamp). This can be ascertained from Eq. (2.10).

2. The realisation of measurements is facilitated, because
   - It is not necessary to use a reference beam to compensate for the changes in radiation power, for the long-term stability of the laser's radiation power is good.
   - The measurement ports may be small due to the parallelism and small diameter of the laser beam. Therefore the flow of the purge air can be smaller and measuring errors remain more insignificant. Often, a separate purge air flow is not needed, for the air flowing inward from the measurement port will be sufficient.
   - The alignment of the laser beam is simple because of the brightness of the beam. This enables the laser and the detector unit to be placed sufficiently far from the measuring channel so that the movements of the channel walls do not disturb the measurement.
   - The laser equipment is small in size. Therefore it can be installed in difficult locations.
   - The radiation power of the laser is high. Therefore higher particle concentrations can be measured than by using an incandescent lamp.
   - When the laser is used as the source of the measuring light, the influence of disturbing gases (e.g. H_2O, CO_2 and CO) disappears in practice. When using an incandescent lamp one has to be certain that the wavelength band of the measuring light does not contain absorption zones of disturbing gases.

2.3.4. Error Sources in Measurement

Error sources in the determination of particle concentrations can be either properties derived from the particulate material or particle behaviour, or disturbances caused by the method of measurement.

Errors due to the properties of particles can be [40]

1. Porosity of the particulate material. Often the particle, such as a particle of fly ash generated in the combustion of oil, is like a porous sphere. Its volatile part has "boiled" away, and sponge-like, porous solid matter is left. The porosity decreases the density of the particle, in which case the mean density to be used in the calculations is generally too high.
2. Coagulation of the particles. The coagulation is a process in which particles stick together forming particle clusters. The estimation of the degree of coagulation is difficult. A particle cluster that has coagulated is observed as one big particle in an optical transmission measurement, and the measurement result is distorted.

Error sources due to the measurement act may be

1. *Influence of the measuring distance.* In practical measurement situations the transmission ratio ($T = I/I_0$) can vary quite a lot. The measurement becomes difficult when measuring the highest particle concentrations, if the

transmission ratio is lower than 10^{-4} or higher than 0.7 [44]. In the first-mentioned case the measuring distance has to be shortened to reduce the error. In the second case it has to be increased. Principles for reducing the measuring distance are represented in Fig. 2.18. Increasing the measuring distance is exemplified in Fig. 2.19, where the distance has been doubled by using a reflecting mirror, or the so-called retroreflector.

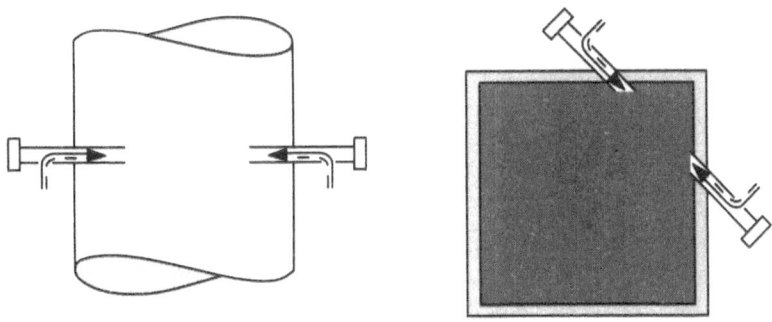

Figure 2.18. Shortening the measuring distance [44].

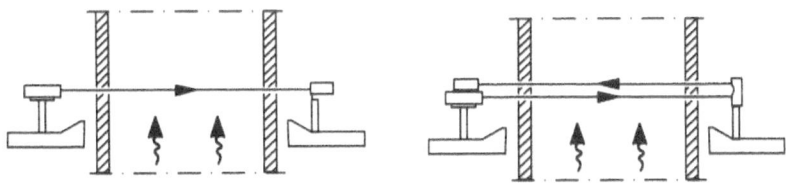

Figure 2.19. Lengthening the measuring distance [44].

2. *Influence of the transmission measured.* The transmission measurement can be distorted, for instance, by the following factors:

- Fouling of the optical components used in the formation of the measurement beam
- Misalignment
- Presence of light-absorbing gas components in the emission gas to be measured, especially if an incandescent lamp is used as the light source
- External factors, such as variations in the operating voltage, moisture, vibrations, corrosive gases, and other factors

In connection with optical measurements, the use of purge air must not be forgotten. If there is overpressure in the measurement channel, for instance, viewports equipped with glass windows have often to be used in order to be

able to direct the laser beam into the channel and away from it [44]. Then
the window has to be purged at least from inside so that the windows stay
clean. The purge air has to be clean and oilless. The purge air system in the
laser measurement of particle concentrations is illustrated in Fig. 2.20.

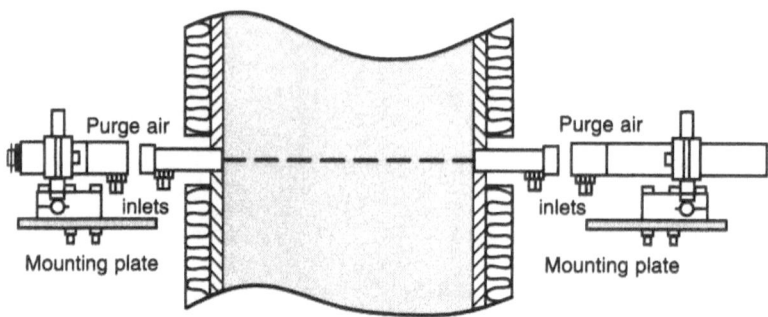

Figure 2.20.   Purge air system [45].

3. *Influence of the variation in the particle size distribution.* The size distribution
   of the particles to be measured can vary during the process so that the
   average radius of the distribution, or its width (represented by the mean
   deviation) change in the course of time. The behaviour of the transmission is
   a very complex process that is difficult to control, as for example in the
   ranges of visible light and near infrared (0.4–1.8 μm). Measuring
   instruments usually utilise this wavelength range. Polydispersive particle
   populations have at least two variable parameters, which determine the
   distribution, namely the mean radius and the standard deviation. This kind
   of a situation is very difficult to control in the practical measurement work.

2.3.5. Reliability of Results of Optical Method

Optical instruments presently available for measuring particle concentrations
operate in the range of Mie scattering. Therefore they cannot utilise the general
equation (2.10) for particle concentration. The operation of the instruments
has, for this reason, to be based on the calibration using some known particle
population. Calibration measurements establish the relationship between the
light attenuation and the particle concentration.

As stated earlier, optical measuring instruments are not capable of taking
into account, for instance, the changes occurring in the particle size
distribution, and hence may yield erroneous results. The calibration cannot
therefore be assumed to be valid for such emission sources, where changes
taking place in the working conditions cause alteration in the particle size
distribution of the emission.

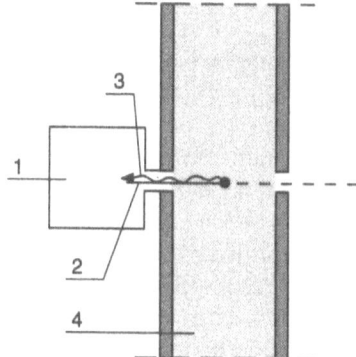

Figure 2.21. An instrument for monitoring of the concentration of particulate material based on the backscattering of laser light. 1, laser particulate meter; 2, laser beam; 3, backscattering to detector; 4, gas-carrying duct.

2.3.6. Method Based on Backscattering of Light

In some cases, a measuring method based on the backscattering of the laser light is useful. Suitable equipment is available on the market which can be used, for example, when a measurement through the channel cannot be arranged, or when the alignment is difficult. Figure 2.21 shows the operating principle of this kind of device. In this arrangement, there is an opening on the opposite side of the measuring channel so that the beam reflecting back from the wall does not disturb the measurement.

Measuring methods based on the scattering of light are functionally more uncertain than those based on light attenuation [40]. This is a conclusion based on both theory and practice. When measuring instruments based on the backscattering of light have been tested, it has been noticed that the result they produce is affected by the quality of the particles, the size distribution, and even the colour of the particles [46].

2.4. Features of Sampling Method and Continuous Methods: Advantages and Disadvantages

The manual gravimetric method is primarily a reference method for the determination of particulate matter emitted from stationary sources and it can also be used for calibrating automatic continuous particulate monitors [35]. This method can be used to determine concentrations ranging from $0.005\,\mathrm{g\,m^3}$ to $10\,\mathrm{g\,m^3}$. The method should be applied as much as possible under steady state conditions of the gas flow in the duct. The manual method is laborious and can take several hours. It cannot be used for continuous indication of particle concentrations or for control purposes. It is considered reliable, but a very fine fraction of the particulate material can still penetrate the filters used for the sample collection.

Continuous monitoring techniques should be applied only on site-specific basis by correlating them with the manual testing method [35]. Continuous techniques offer advantages whenever a quick indication of particulate emissions or monitoring of variations in the emission levels is needed. Continuous particulate monitoring signals can be used, e.g. for process control purposes.

Continuous particulate monitors must be calibrated. The calibration must be repeated if the conditions (emission controls, fuel type etc.) change. The mass concentration range of continuous monitors is applicable only when calibration specifications are met. A continuous method is not applicable if changes in physical properties (size, shape, colour etc.) or in chemical composition of the particulate matter impair the integrity of the calibration.

When carefully calibrated, the continuous optical method, for instance, can be reliably used in certain applications [40], bearing in mind the following:

- The measuring instrument should be calibrated for each particulate material separately, so that the changes in the optical properties of the particulate material can be eliminated when moving from one type of emission source to another.
- While measuring relative particle concentrations, the measurement results are comparable when the optical properties and the particle size distribution of the particulate material remain constant during the measuring period.
- When measuring absolute particle concentrations, the measuring equipment has to be calibrated for the particulate material to be measured and for each size distribution separately. During the measuring period these have to be the same as during the calibration.
- It is easier to calibrate a measuring instrument using an incandescent lamp as light source than a laser-based device. The calibration of an instrument based on the use of an incandescent lamp is not nearly as sensitive to errors caused by variations in the particle size distribution, for example, as the calibration of an instrument based on the use of laser.

Optical measuring instruments can be used for certain applications fully reliably, provided that the operating personnel has the expertise. Possible applications include

1. Control of processes: drying, combustion (e.g. soda recovery boiler) and dust separation (e.g. control of the condition of filters).
2. Control of emissions and relative emission measurements.

Chapter 3

Determination of Particle Size Distribution

3.1. Introduction

It is important to know the size distribution of the particulate material in emission gases when, for instance, developing control technologies and testing their efficiencies, as well as when monitoring emissions. The size and the size distribution of solid particles can be measured by various methods. The choice of method is affected by the particle size, for example. The measuring methods vary from mechanical, inertia-based separators (e.g. impactors and cyclones) to methods exploiting optical and electrical interactions. In the following, the most commonly used of these are discussed.

3.1.1. Aerodynamic Diameter of Particle

The aerodynamic diameter of a particle is defined as that of a sphere whose density is $1 \, \mathrm{g \, cm^{-3}}$ (cf. density of water) and which settles in calm air at the same velocity as the particle in question [2]. We are dealing with this diameter when considering, for example, collection efficiencies, and it is the result when dynamic classifiers (e.g. cascade impactors) perform size classifications.

3.1.2. Curvilinear Motion of Particle

When a flowing gas meets an obstacle, the direction of its motion changes. This means that the particles travelling with the gas are also forced into curvilinear motion. The properties of the particles, above all, their mass, influences how well the particles are capable of following the changes in the direction of the gas flow.

The motion of solid particles in the gas flow that changes its direction, is usually characterized by using a few characteristic parameters. These include the *stopping distance* and the impaction parameter or the *Stokes number*.

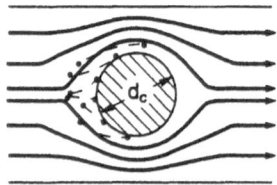

Figure 3.1. The motion of the particles when the gas flow meets a cylindrical obstacle [2].

The stopping distance S is the greatest distance that a particle moving at a definite velocity can travel in calm, unobstructed air when it is not influenced by external forces:

$$S = V_0 \tau \tag{3.1}$$

where V_0 is the initial velocity of the particle, and τ is the relaxation time of the particle. The relaxation time characterises the time that the particle needs to condition (or, "to relax") its velocity to the new situation of acting forces [47].

The Stokes number is a unitless number which characterises the curvilinear motion of a particle and which connects the stopping distance of the particle with a dimension representing the size of the stationary obstacle. For example, for a cylinder-shaped obstacle (Fig. 3.1), whose diameter is d_c, the Stokes number (St) is

$$St = \frac{S}{d_c} = \frac{\tau U_0}{d_c} \tag{3.2}$$

where U_0 is the disturbance-free velocity of the gas far from the cylinder.

When the Stokes number is high, the particles strongly resist changing their direction with the change of direction of the streamlines of the gas flow. When the Stokes number approaches zero, the particles follow the streamlines of the gas flow completely.

3.1.3. Inertial Impaction

Impaction is based on the fact that the path of a particle is changed by the action of an obstacle. This phenomenon is used for collecting and measuring particles. Through the past three decades, the phenomenon has generally been applied in cascade impactors.

All inertial impactors have a common operation principle [47]: the particle is carried through a nozzle with a gas stream, which is directed against a plate (Fig. 3.2). This plate, the impaction plate or the collection plate, changes the flow so that it develops an abrupt 90° turn. Particles which are too slow are not capable of following the change in the direction of flow, and collide with the plate. The cross-section of the impactor is represented in Fig. 3.2.

The collection efficiency of an impactor for a monodispersive particle population (formed of particles having the same aerodynamic diameter) can be estimated by means of a simplified model shown in Fig. 3.3. It represents half

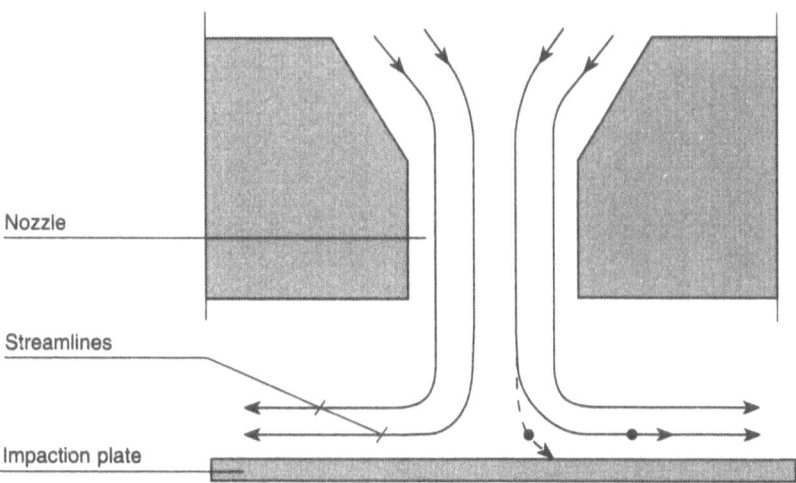

Nozzle

Streamlines

Impaction plate

Figure 3.2. Cross-section of an impactor. Light particles follow the streamlines of the impactor, but heavy particles (dashed line) collide with the impaction plate [47].

of a rectangular impactor. There the velocity of the particle in the direction of the radius r is v_r, based on the centrifugal force. Let us assume that the flow rate at the tip of the nozzle is even, and that the streamlines curve along the circumference of a circle of centre A. Then the particles which are closer to the centre line than the distance Δ are retained in the impaction plate. This means that the resulting collection efficiency is the proportion of particles retained in the impaction plate, which is equivalent to Δ/h. In order to examine the collection efficiency, a test series would have to be performed using various particle sizes.

The impaction theory is used to explain the dependence of the collection efficiency on particle size [47]. Figure 3.4 shows a typical collection efficiency curve for an impactor. The Stokes number, or the impaction parameter, characterises the collection efficiency. For an impactor the Stokes number is determined as a ratio of the stopping distance of the particle to the radius of the nozzle. The velocity of the particle used here is the average exit velocity (U) from the nozzle:

$$\text{St} = \frac{\tau U}{D_j/2} \tag{3.3}$$

where D_j is the diameter of the nozzle.

The Stokes number, thus defined, differs slightly from the previous definition. Here the dimension of the nozzle is used instead of an obstacle-related dimension. A definition like this has proved practicable, since when

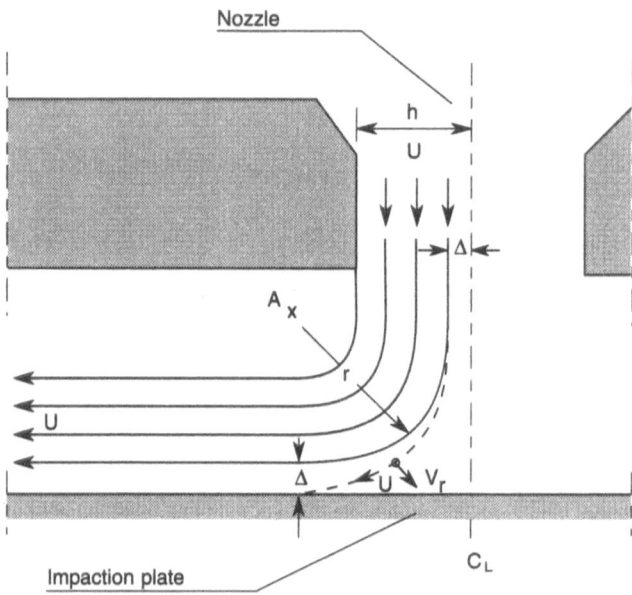

Figure 3.3. Simplified model for the estimation of the collection efficiency for monodispersive particles. The particles which are closer to the centre line of the nozzle than $\varDelta$ are retained in the impaction plate [47].

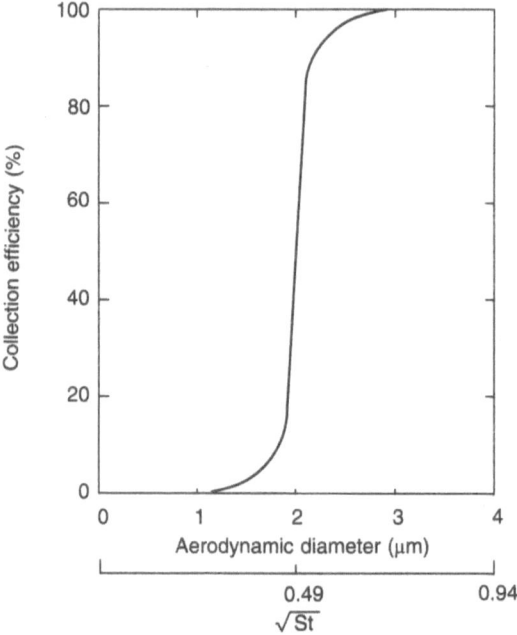

Figure 3.4. Typical collection efficiency of an impactor [47].

estimating the operation and collection efficiency of an impactor, the dimension of the impactor nozzle is a more essential parameter than, for instance, the distance of the collection plate from the nozzle, even though that is also significant.

The Stokes number characterises the collection efficiency of an impactor. It links a property related to the particle and a property of the impactor to one another. The Stokes number is proportional to the second power of the particle diameter. This can be seen from Fig. 3.4, in which the collection efficiency has been represented as a function of both the particle diameter and the square root of the Stokes number.

The mass of the particles collected in the impaction plate can be determined by weighing. An impactor can be used, for example, so that the particles passed through it are collected on a filter. Thus the particle population can be divided into two parts. If, for instance, the cut diameter of an impactor is 5 μm, it may be noted that 30% of the total particle mass are particles having a larger aerodynamic diameter than the cut diameter, and 70% having a smaller diameter [47].

From the cumulative mass distribution several points can be determined, if various flow rates with corresponding different cut diameters are used in the impactor. This method can encounter practical difficulties, as one cannot always be certain that the size distribution of the particle sample entering the impactor remains the same and representative when using various flow rates. This drawback can be eliminated by using simultaneously several impactors with different cut diameters. Impactors are normally used in series, since then the control of many flow rates is avoided, which would have to be performed for parallel impactors. A group of series-connected impactors is called a *cascade* impactor.

3.2. Cascade Impactor

A cascade impactor, which can be placed inside a gas channel, is a simple and practicable device for the estimation of particle size distribution [48]. The sample can be collected isokinetically, and the device can withstand also the conditions of a flue gas.

A cyclone, for instance, can be connected to the front part of a cascade impactor to collect coarser particles ($>2.5\,\mu$m). At the end, a filter is often mounted to collect the fine fraction. Figure 3.5 represents the structure of a cascade impactor.

Individual impactors of a cascade impactor are called impactor stages. These are arranged in order with the largest cut diameter being first and the smallest last. The cut diameter is reduced stage by stage by reducing the size of the nozzle or the number of nozzles, or the distance of the impaction plate from the nozzle. The impaction plates are demountable, so that the collected

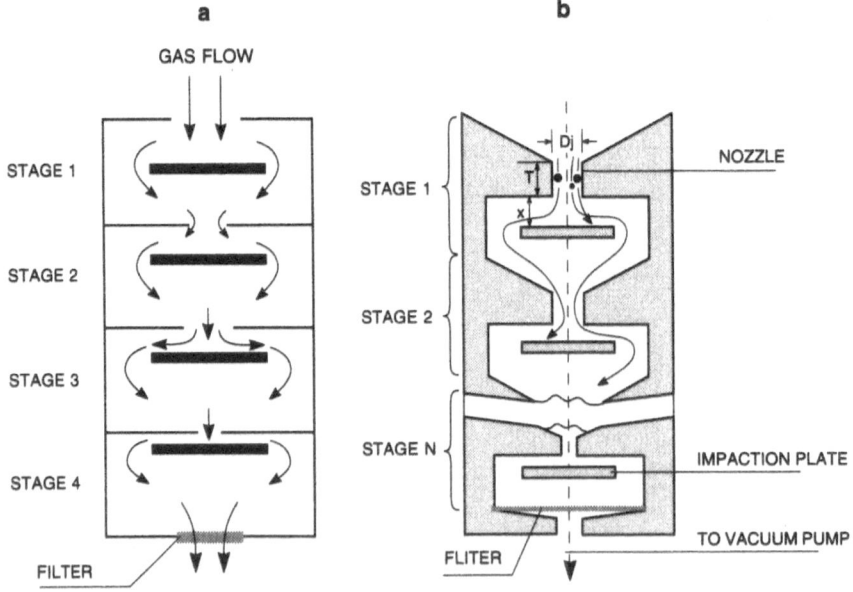

Figure 3.5. **a** Operation principle and **b** cross-section of a single-jet cascade impactor [2,49].

particles can be weighed or analysed. After the last stage, there is usually still a filter to collect particles smaller than the cut diameter of the last stage.

Each stage of a cascade impactor is assumed to collect all the particles entering it which are larger than its cut diameter. The cumulative mass of the particles is normally plotted in the graph of the particle size distribution as a function of the upper limit of the particle size range corresponding to each stage. By means of a cascade impactor, particle samples can even be classified into 12 fractions.

The theory of particle separation based on inertia has been developed and tested for several decades. Theoretical and experimental research has been carried out in order to improve the collection efficiency of the impactor and to develop design factors affecting the efficiency.

3.3. Cascade Centripeter Collector

Particle sizes can also be classified using a cascade centripeter collector. Particles are divided into five size categories. In a cascade centripeter collector the coarsest particles are collected in the first filter, and the finest in the filter at the bottom. The operation principle of the cascade centripeter collector and its cross-section are shown in Fig. 3.6. In the collector, the fine particles travel

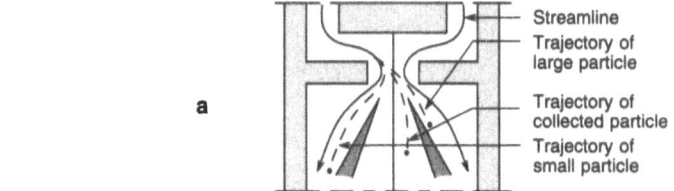

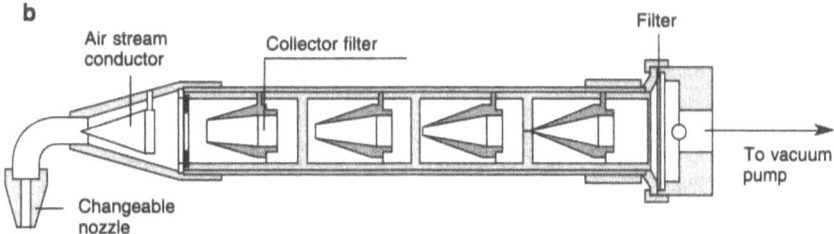

Figure 3.6 **a** Operation principle and **b** cross-section of the cascade centripeter collector [50].

with the gas flow, whereas the coarse particles in the middle of the gas flow go into the nozzle and are collected on the collecting filter at the front end.

3.4. Multistage Cyclone

In a multistage cyclone an isokinetically sampled flue gas meets the cyclone body tangentially, and generates a turbulent flow [51, 52]. Particles larger than the cut diameter move radially towards the wall of the cyclone body by the action of the centrifugal force to settle finally in the collector. Particles smaller than the cut diameter are returned back and pass through the outlet pipe into another cyclone which has a smaller diameter. This collects its fraction of the particles. In a multistage cyclone there are several cyclones (e.g. six) in series. After the cyclones, there is a filter which collects particles having a smaller diameter than the cut diameter of the last cyclone. The operation principle of a cyclone is shown in Fig. 3.7. A multistage cyclone can be used to classify particles ranging from 0.5 to 20 μm in diameter.

3.5. Aerodynamic Particle Sizer

An aerodynamic particle sizer can be used to classify particles whose aerodynamic diameter is 0.5–30 μm [51]. In a constant gas flow particles reach a velocity dependent on their aerodynamic diameter. The velocity for each particle is determined optically.

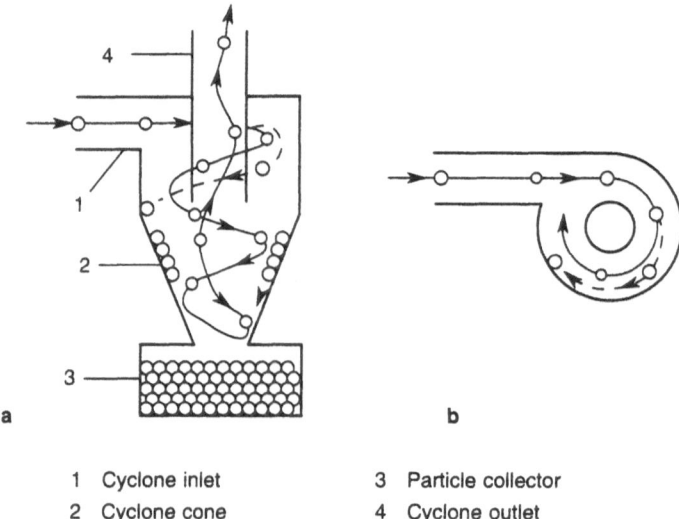

1 Cyclone inlet	3 Particle collector
2 Cyclone cone	4 Cyclone outlet

Figure 3.7. Cross-section showing the operation principle of a cyclone **a** from the side and **b** from above [51].

3.6. Analysis of Particle Size Distribution Based on Fraunhofer Diffraction of Laser Light

3.6.1. Origin of Diffraction

When a beam of light is incident on an obstacle, some specific effects are noted at the edges of the obstacle. These effects are due to the optical differences of the media meeting at the edge, where there is an abrupt change from one medium to another with regard to the propagation of light. In our case, when solid particles in a gas are considered, one of the media can be thought of as fully transparent, whereas the other is practically opaque. The result of this discrete change in the properties of the media at the edge is the change of the direction of some amount of light incident at the edge. This means "bending" of part of the light around the edge, and in practice, this effect can be noticed as a smoothing of the edges of the image or shadow of a piece as illuminated by a light source.

When one of the different media for the propagation of electromagnetic radiation, such as light or X-rays, is contained in another and forms only a small portion of the whole, as for instance particles in a gas or atoms in a volume of solid matter, a phenomenon called *diffraction* occurs. It is a disturbance in the propagation of the wave. An obstacle in a transparent medium, or an aperture in an opaque object, can be referred to as a *diffraction centre* [53]. Diffraction effects are most pronounced when the wavelengths of the waves involved are comparable with the dimensions of the diffraction centres. Particularly striking effects are observed, when large numbers of identical diffracting objects are arranged in a regular way. For these reasons X-ray diffraction is used for the research on the structures of solids. In this context we consider the diffraction of light, which can be used for the determination of the particle size distributions of particles in emission measurements.

Diffraction of light originates from the vibrations caused by the incident light in parts of the diffraction centres. These vibrations are also the origin of the scattering, the reflection and the refraction of light. The *forced* vibration of the electric charges in the molecule caused by a light wave at the frequency of the incident light generates new waves of the same frequency. Diffracted waves can interfere with each other and with the original wave forming diffracted waves as a superposition.

3.6.2. Huygens' Principle

Huygens' principle states that every point of a propagating wave front can be considered as the source of a secondary wave. This is a geometrical construction which can be used to visualise interference effects of the secondary wavelets. It is illustrated in Fig. 3.8 with a plane wave meeting a strip-like obstacle perpendicular to the plane of the paper. We consider a case where the width of the obstacle is many times the wavelength of the light, and can notice that the resulting wave front is parallel to the incident light except at the edges of the obstacle. Here the diffraction causes the light wave to "bend", as indicated by two tangents of constructive interference, so that some light is reaching the area that should be in shadow in a strict geometrical consideration (Fig. 3.8) [54]. A diverging wave results.

3.6.3. Fraunhofer Conditions of Diffraction

Superposition of light can be observed in certain directions beyond the diffraction screen or diffracting obstacles. In our case of the determination of particle sizes it is supposed that the observations are made at a long distance from the diffraction centres. This distance is orders of magnitude longer than the dimensions characterising the particles being measured. The light beams from the object studied can thus be considered to be parallel at the observation point. The approximation becomes precise only at an infinite distance from the diffracting screen and the *Fraunhofer conditions* of diffraction are valid.

Fraunhofer diffraction is of importance in practice and in the special case of the measurement of particle sizes. This is because the conditions of the observations at infinity can be realised using a lens. The image of the pattern at infinity is brought into the focal plane of the lens. The lens used in the particle size analysers is sometimes called a *Fourier lens*, as the complex amplitude of

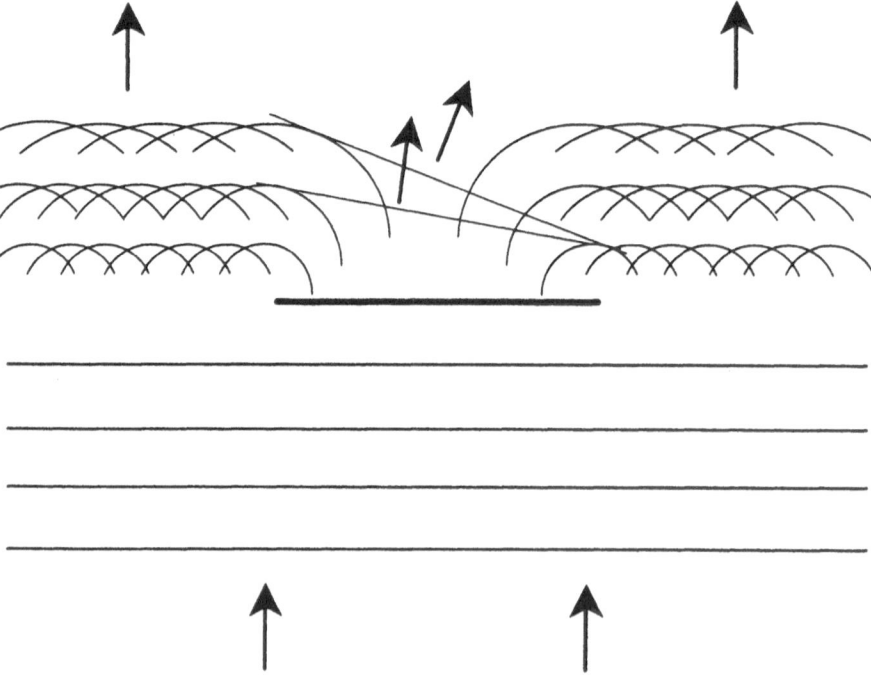

Figure 3.8.   Huygens' construction for the propagation of light meeting an obstacle.

the diffracted disturbance on its focal plane, caused by the obstacles, is the Fourier transform of the configuration formed by the obstacles.

3.6.4. Interference

A pattern of darker and brighter zones can be observed on a screen, when plane waves of light are allowed to pass through a set of obstacles. This is due to the interference of the secondary waves originating from the edges of the obstacles (Fig. 3.9). When making observations at a long distance, we could approximate this distance of observation to be the same from both edges of each obstacle. The path difference must, however, be taken into account when the *phase difference* between waves originating from different edges of the obstacle are calculated. In this connection the path difference is of crucial importance. This is because we are dealing with path differences which are in the range of the wavelength of the incident light, even though this difference is negligible as compared to the observation distance. Phase differences between waves from different edges of the obstacles (here assumed monosized) determine the conditions of *interference*, as illustrated in Fig. 3.9. On the screen, a pattern of brighter and darker zones will be formed. Electronic

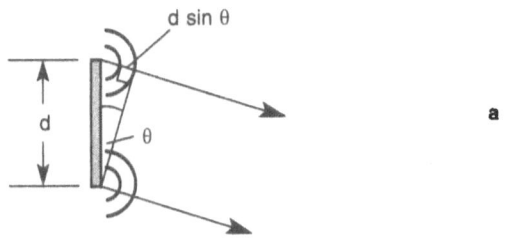

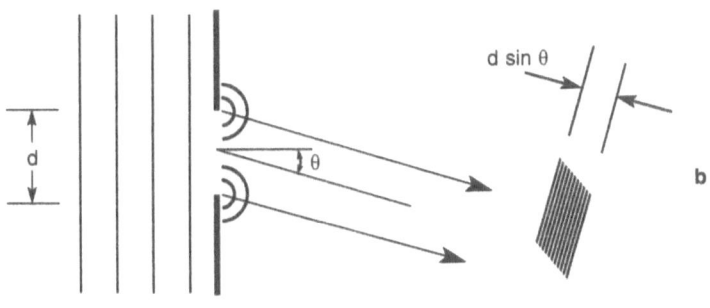

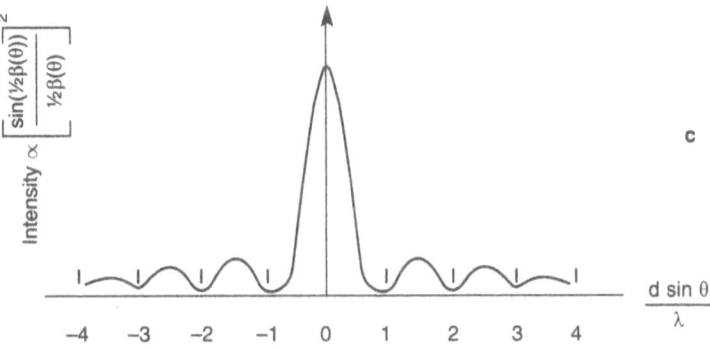

Figure 3.9 **a** Interference of wavelets. **b** Single-slit diffraction [53] and **c** the resulting diffraction pattern.

vibrations at the edges of the strips induced by incident light generate waves that interfere to form the light pattern.

The waves will interfere constructively, when the path difference between them is an integer multiple of the wavelength of the light. This corresponds to certain values of the angle θ between the incident and diffracted light and is determined by the relation

$$\sin \theta = \frac{n\lambda}{d} \tag{3.4}$$

as can be deduced from Fig. 3.9. Here d is the width of the obstacle, λ is the wavelength of the light, and $n\lambda$ is an integral number of wavelengths. At these angles, *bright lines* appear on the screen. Between these angles are the values of angles at which destructive interference occurs, causing dark lines on the screen. The angles of the constructive and destructive interference depend on the size of the obstacles. As can be concluded from the equation above, narrower obstacles cause corresponding interference zones at larger angles than do wider obstacles.

From Fig. 3.9 it is noticed that the path difference between the two waves shown is $d\sin\theta$. The corresponding phase difference is $kd\sin\theta$, where k is the magnitude of the wave vector.

3.6.5. Single-Slit Diffraction

For a while we shall consider an aperture (Fig. 3.9(b), a uniform slit perpendicular to the plane of the paper) rather than an opaque obstacle, as this is a simple case and results in a diffraction pattern very similar to that from monosized spherical particles.

If the amplitude of the disturbance from each imaginary wavelet centre to direction θ is $A(\theta)$ and $A_{max} = A(0)$, it can be shown by simple geometrical consideration that

$$A(\theta) = \frac{A_{max}|\sin\frac{1}{2}\beta(\theta)|}{\frac{1}{2}|\beta(\theta)|} \tag{3.5}$$

where $\beta(\theta) = kd\sin\theta$ is the phase difference between the disturbances originating from the two edges of the slit [53].

The variation of intensity with direction can thus be indicated by plotting $(A/A_{max})^2$. This is shown in Fig. 3.9(c). Most of the wave energy is concentrated into the forward maximum at $\theta = 0$, though there are also other small maxima.

In the case of particles contained in a transparent medium, such as air, the diffraction pattern on the screen or the detector plane is composed of lighter and darker rings or the pattern is radially symmetrical. In practical applications, particles are of different sizes, forming a *size distribution*. The diffraction patterns of individual particles are superimposed, which makes the rings diffuse rather than sharply separated.

3.6.6. Diffraction Pattern of Circular Obstacles

In the particle size analysis we use the diffraction pattern at angles near the direction of the incident light. When the angle and the dimensions in both the object space (particles) and at the image or detector plane are essential in the analysis, the case is readily treated in polar coordinates [55].

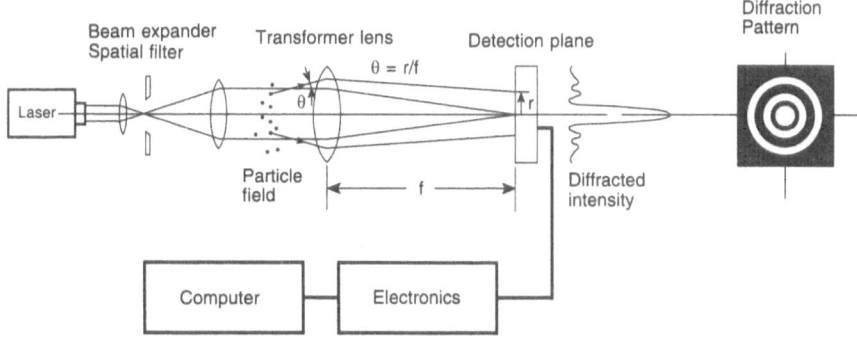

Figure 3.10. Scheme of particle size analysis based on the diffraction of forward-scattered laser light.

We now consider particles illuminated by a collimated laser beam as in Fig. 3.10 [56]. A monodispersive ensemble of spherical particles that are large compared with the wavelength would produce the characteristic Airy diffraction pattern (neglecting anomalous diffraction), as described by Fraunhofer diffraction theory:

$$I(\theta) = cI_{inc} \frac{\pi^2 d^4}{16\lambda^2} \left[\frac{2J_1(\pi d\theta/\lambda)}{\pi d\theta/\lambda} \right]^2 \tag{3.6}$$

where $I(\theta)$ is the scattered intensity at the angle θ measured from the laser beam axis, c is a proportionality constant, I_{inc} is the intensity of the incident beam, J_1 is the first-order Bessel function of the first kind, d is the particle diameter, and λ is the laser wavelength. The small angle approximation of $\sin \theta = \theta$ has been made in Eq. (3.6). The coefficient of the bracketed squared term is the on-axis scattering intensity $I(0)$. The form of the Airy diffraction pattern is shown in Fig. 3.11, and it can be seen to be very similar to the diffraction pattern shown in Fig. 3.9(c).

In practical systems a distribution of particle sizes, or polydispersion, is generally encountered. The composite scattered intensity profile is a linear combination of the characteristic profiles of each particle size with a weighting coefficient equal to the number of particles of that size in the sample volume. The diffraction signature of a polydispersive spray is given by

$$I(\theta) = cI_{inc} \int_0^\infty \frac{\pi^2 d^4}{16\lambda^2} \left[\frac{2J_1(\pi d\theta/\lambda)}{(\pi d\theta/\lambda)} \right]^2 n(d) \mathrm{d}d \tag{3.7}$$

where n is a differential number distribution such that $n(d)\mathrm{d}d$ is the number of particles in the laser beam with sizes between d and $d + \mathrm{d}d$ [56]. Equation (3.7) assumes a uniform intensity profile across the laser beam (constant I_{inc}). $n(d)$ integrated over all particle sizes is the total number of particles in the beam.

The scattered light that is refracted by the receiving lens (collecting lens, transform or Fourier lens) in Fig. 3.10 is directed onto the transform plane

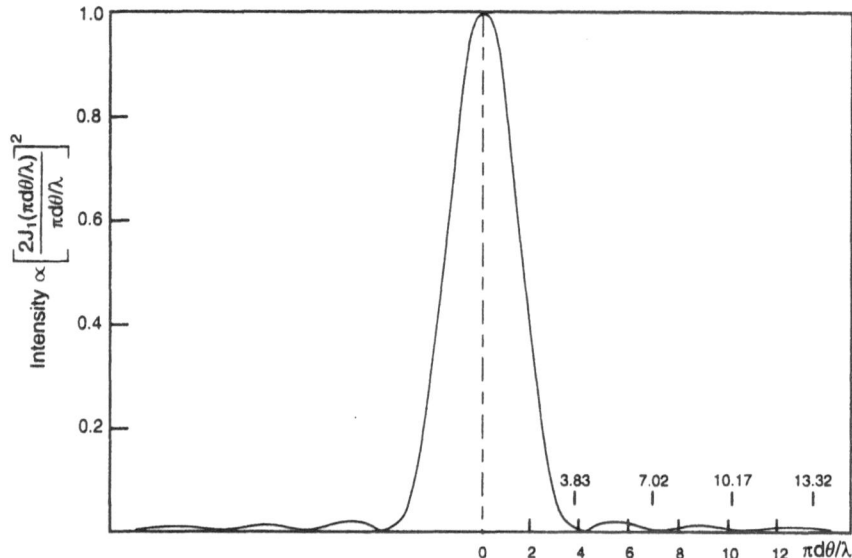

Figure 3.11. Diffraction by spherical monosized particles [57]. Variation of intensity in the focal plane along each axis.

(detector plane, focal plane of the lens) at radial positions given by (neglecting lens aberrations)

$$r = f\theta \tag{3.8}$$

where f, r and θ are defined in Fig. 3.10. Note that Eq. (3.8) is independent of scattering particle position. This means that the sample particles can be stationary or move within the sample volume. They can also be suspended in a gas or in a liquid.

Consider an array of annular detector elements where r_{ij} and r_{oj} are the inner and outer radii, respectively, of the jth ring detector. The jth detector collects (neglecting vignetting by the receiving lens aperture) a hollow cone of scattered light defined by inner and outer scattering angles θ_{ij} and θ_{oj}, which are related to r through Equation (3.8). The scattered light energy S_j collected by the jth finite aperture annular ring detector is obtained by integrating $I(\theta)$ from Equation (3.7) over the aperture, giving [57]

$$S_j = cI_{inc} \int_0^{\infty} \frac{\pi d^4}{4} \left(J_{0ij}^2 + J_{1ij}^2 - J_{0oj}^2 - J_{1oj}^2 \right) n(d) \mathrm{d}d \tag{3.9}$$

where

$$J_{npj}^2 = J_n^2 \frac{\pi d r_{pj}}{\lambda f}$$

with n indicating the order of the Bessel function and p indicating inner (i) or outer (o) detector radius.

3.6.7. Inversion of Scattering Data

The determination of the particle size distribution $n(d)$ from measured light-scattering signatures $I(\theta)$ or S_j is done by the inversion of the scattering data. Integral transform inversions of Eq. (3.7) are possible [58], but are not commonly used at present. Generally, the size distribution is divided into a finite number of discrete size classes, typically of the order 2^3 or 2^4 [56]. In some cases, the size distribution function is assumed to follow some common form with two degrees of freedom, such as Rosin–Rammler [57] or log-normal. Other approaches do not constrain the results to a particular form and are termed model-independent inversion methods [59].

For monosized circular or spherical particles, the radius r_0 of the first or the smallest dark ring can be calculated from the equation (from Fig. 3.11, $3.83/\pi = 1.22$)

$$r_0 = 1.22 \frac{\lambda f}{d} \tag{3.10}$$

The smaller the diameter of the particles d the bigger is the radius r_0.

3.6.8. Useful Range of Particle Sizes

It was assumed earlier that the diameter of the particles or the obstacles is greater than the wavelength. This is the range of geometrical optics, and the diffraction pattern according to Fraunhofer conditions of diffraction appears in the focal plane of the lens. The system operates at angles close to the direction of the incident monochromatic light.

Diffraction influence begins to form maxima and minima in the scattering pattern around the direction of the forward beam already in the Mie scattering range with particle diameters smaller than the wavelength, if we imagine the particle diameter increasing. The wavelength of 632.8 nm of the He–Ne laser is generally used in laser particle size analysers. It is therefore possible to carry out measurements of size distribution even in the boundary area of the Mie scattering range. Practice has shown that measurements with a laser diffraction spectrometer of particles with a size as small as 0.2 μm can be correct and reproducible.

The measurement range can go up to about 1000 μm at the highest. Measurement units can use detectors with 30 channels to obtain high resolution.

3.6.9. Construction of Analyser

The particle size analyser based on the diffraction of laser light consists of laser source, beam expander, collector lens and detector (Fig. 3.10). The detector contains light diodes arranged to form a radial diode array detector. The particle sample to be measured can be blown across the the laser beam (dry sample) or it can be circulated via a measurement cell in a liquid suspension. In the latter case, the laser beam is directed through the transparent cell. There are also constructions which allow the operation of the analyser for (practically) continuous monitoring and process control. In some constructions, laser source and receiver units can be introduced into high temperature environments inside a water-cooled probe [60]. Particles flow through a flow access region of the probe, and the light signal is transferred to the detectors by fibre optics.

3.7. Determination of Size Distribution of Very Small Particles

The dynamics of aerosols (condensation, evaporation, coagulation or deposition) depend on the particle size [61]. Aerosols are small particles dispersed in gas. In air, for example, they form a dilute suspension of relatively immobile, massive particles (compared with gaseous species) [62]. Table 3.1 compares properties typical of a nitrogen gas molecule and an aerosol particle.

Table 3.1. Size, mass and concentration of aerosol particles compared with gas molecules [62]

	Diameter (μm)	Mass (g)	Concentration (cm^{-3})
Gas molecules (e.g. N$_2$) (e.g. N$_2$)	0.00038	4.6×10^{-23}	$\approx 10^{19}$
Aerosols	0.01–10	10^{-18}–10^{-9}	$< 10^8$

Within the aerosol particle size range of 0.01–1.0 μm, the size distribution can be determined based on the electrical mobility of particles [61]. The electrical mobility depends on the viscosity of the gas and on the size and the charge number of the particle. Particles smaller than this size range can be calculated by using a condensation nucleus counter, if the concentration of particles is small.

Aerosols with a particle size range of 0.01–1.0 μm are an important research area of aerosol physics and chemistry, as in flue gases for example, the concentration of heavy metals and PAH compounds is highest on small particles. On the other hand, particle collectors, such as fabric filters, are unable to retain particles of this size range. Metals and PAH compounds can thus easily migrate into the environment as well as into the respiratory organs of people and animals.

The measuring method based on the electrical mobility has a high resolution, but its operation conditions are limited. The temperature of the aerosol must be near $+25\,°C$ and the concentration smaller than 10^7 particles per cubic centimetre. A diluting unit is needed in order to be able to measure the particle size distribution from a stack flue by means of the electrical differential mobility analyser.

3.7.1. Differential Mobility Particle Sizer

The differential mobility particle sizer (DMPS) measures, as mentioned, the size distribution of aerosols on the basis of electrical mobility in the particle size range $0.01–1.0\,\mu m$. The use of the device provides the removal of larger particles prior to the analysis.

In measurement based on electrical mobility, aerosol particles are classified according to size by means of an electrical classifier [61]. The number of particles is counted by a condensation nucleus counter (CNC). A micro-computer controls the equipment, collects the data to be processed, and computes on that basis the concentrations of particles with different diameters. The computer prints out the concentrations classified according to particle size. The operation principle of the equipment is shown in Fig. 3.12.

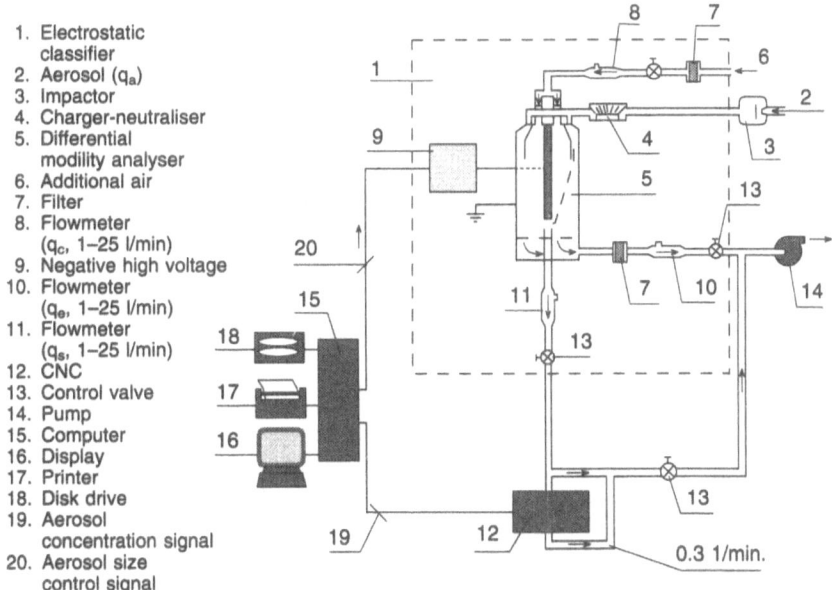

1. Electrostatic classifier
2. Aerosol (q_a)
3. Impactor
4. Charger-neutraliser
5. Differential modility analyser
6. Additional air
7. Filter
8. Flowmeter (q_c, 1–25 l/min)
9. Negative high voltage
10. Flowmeter (q_e, 1–25 l/min)
11. Flowmeter (q_s, 1–25 l/min)
12. CNC
13. Control valve
14. Pump
15. Computer
16. Display
17. Printer
18. Disk drive
19. Aerosol concentration signal
20. Aerosol size control signal

Figure 3.12. Operation principle of the measuring device based on electrical mobility. In the figure, q_a denotes the volumetric flow of the ingoing aerosol, q_c that of the carrying air, q_s the flow of the exiting monodispersive aerosol, and q_e the volumetric flow of the aerosol to be removed [63].

The determination of size distribution by means of an instrument based on electrical mobility consists of two stages [64]:

1. Measurement of the mobility distribution (by gradually increasing the centre electrode voltage).
2. Calculation of the size distribution of particles (the proportion of charged particles and the counting efficiency of the condensation nucleus counter must be taken into account).

Operation Principle

The aerosol particles are classified electrically [61]. This is done using a differential mobility analyser, which separates the particles into a sample belonging to a definite range of electrical mobility, which is determined by the centre electrode voltage of the analyser. The differential mobility analyser consists of two coaxial cylinder electrodes, the outer one of which is earthed. The centre electrode is at a negative potential of 0–10 kV relative to the outer electrode. Between these two electrodes, the clean air surrounding the centre electrode and the aerosol particles which form a cylinder around it flow down at the same velocity. The positively charged particles travel though the clean air layer towards the negatively charged centre electrode. Their trajectory depends on the volumetric flow, the geometry of the analyser, the electric field, the diameter of the particle, and the charge of the particle. Only those particles that belong to the selected mobility range (according to the potential difference of the electrodes) pass through the hole near the bottom of the centre electrode. These particles form the aerosol sample. Other positive particles adhere to the surface of the centre electrode, or leave the analyser along with the exit air. The negatively charged particles adhere to the surface of the outer electrode. Those particles which do not pass through the hole are not included in the measurement.

3.7.2. Charging of Particles by Means of Bipolar Diffusion Charger

Aerosol particles are charged and discharged using a radioactive source ^{85}Kr [61]. The degree of charging is determined by factor Nt, where N is the number concentration of the particles, and t is the neutralisation time, or the time the aerosol remains in the region of the influence of the charger.

The uncharged particles need time τ to reach equilibrium

$$\tau = \frac{1}{4}\pi e N Z_i \tag{3.11}$$

where e is the unit charge, N is the number concentration of ions, and Z_i is the electrical mobility of the ions:

$$Z_i = \frac{D_i e}{kT} \tag{3.12}$$

where D_i is the diffusion coefficient of the ion molecule, k is the Boltzmann constant and T is the absolute temperature. When the particle is larger than the mean free path of the ion (the distance the ion travels on average between collisions), τ also represents the neutralisation of a charged particle. The aerosol must remain under the influence of the charger for at least time τ.

The operation period of the ^{85}Kr source is long, for its half-life is 10.3 years, and it can be used for longer than the half-life. The maximum energy of the β (electron) radiation available from this source is only 0.695 MeV. The propagation of this kind of radiation can be prevented by using an aluminium sheet which is 1.6 mm in thickness. ^{85}Kr is also a chemically inert material, which means that it is safer to use than many other radioactive materials.

The charging depends on the activity of the source and the volumetric flow of the aerosol. Minimum values can be derived for the factor Nt according to particle size so that equilibrium can be achieved [65]:

1. When the radius of the particle is much larger than the mean free path of the ion, which is dependent on the particle concentration and which in one experiment was estimated to be 10^{-8} metres at 25 °C, Nt has to be 6×10^6 ions cm^{-3} s.
2. When the radius of the particle is comparable with the mean free path of the ion or less than that, Nt has to be 10^5 ions cm^{-3} s.

When in one experiment a flow rate of 31min^{-1} was used in the particle size range 0.02–0.2 μm, the aerosol remained 28 s in the discharger. The value for the parameter Nt is then 4.8×10^7 ions cm^{-3} s [66]. This value is approximately eight times the value needed, or 6×10^6 ions cm^{-3} s so that the equilibrium is definitely achieved.

3.7.3. Transfer Function

The flow rates of the differential mobility analyser as well as the centre electrode voltage are adjusted so that aerosol of a definite size exits from the analyser [61]. The flow rate of the ingoing aerosol can be adjusted to be the same as that of the outgoing monodispersive aerosol, in which case only two flow indicators are needed. The mean value Z_{pc} for the distribution of the electrical mobility of the aerosol coming out of the differential mobility analyser and the width DZ_p of the mobility channel can be expressed as follows [67]:

$$Z_{pc} = \frac{q_c}{2\pi L V} \ln \frac{R_1}{R_2} \tag{3.13}$$

$$DZ_p = \frac{q_a + q_s}{4\pi L V} \ln \frac{R_1}{R_2} \tag{3.14}$$

where q_c is the volumetric flow rate for the carrying air, R_1 is the radius of the outer electrode, and R_2 is the radius of the centre electrode. V is the centre electrode voltage, and L is the distance between the outlet of the sample aerosol and the inlet port of the aerosol.

On the other hand, electrical mobility is of the form

$$Z_{pc} = \frac{n_p e C}{3\pi\mu D_p} \tag{3.15}$$

where D_p is the diameter of the particle, n_p is the charge number of the particle, μ is the viscosity of the gas, and C is the Cunningham slip correction factor.

$$C = 1 + 2.492\frac{\lambda}{D_p} + 0.84\frac{\lambda}{D_p}\exp\left(-0.435\frac{D_p}{\lambda}\right) \tag{3.16}$$

where λ is the mean free path of the gas molecule. This gives the correlation of the particle size with the centre electrode voltage, the charge number of the particle, the flow rates, and the geometry of the analyser:

$$\frac{D_p}{C} = \frac{2n_p e V L}{3\mu q_c \ln(R_1/R_2)} \tag{3.17}$$

By means of this transformation function it is possible to calculate the diameters of the particles in the sample aerosol leaving the differential mobility analyser, if the charge numbers of the particles are known. The function is the probability that a particle with mobility Z_{pc}, which enters the analyser, exits the analyser as sample aerosol.

3.7.4. Condensation Nucleus Counter

The number of the particles exiting the differential mobility analyser is determined by using a condensation nucleus counter. The structural diagram of this counter is shown in Fig. 3.13. The device consists of a saturation chamber, condensation tube, light scattering indicator, mass flow meter, and a pump. The aerosol is saturated with alcohol (butanol) at the temperature of $35\pm0.1\,°C$. The saturated aerosol is conducted into the condensation tube, where the temperature is $10.0\pm0.1\,°C$. The alcohol is condensed on the surface of the aerosol particles, and the particles grow thereby larger drops of approximately $12\,\mu m$ in size.

The indication of particles is performed by means of white light, lenses, a narrow slit, and a light detector. The white light is focused by means of lenses on the slit which is $0.1 \times 2\,mm^2$ in size. An image of the slit of the same size is formed above the aerosol outlet hole in the condensation tube. The condenser

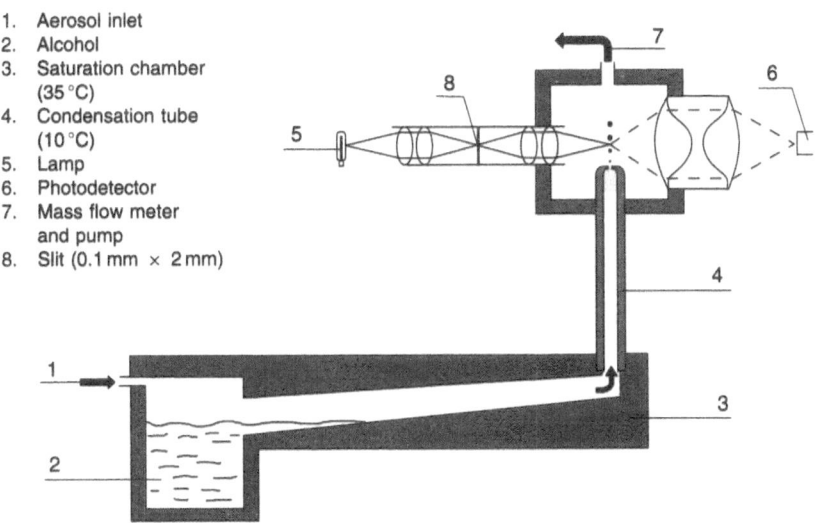

1. Aerosol inlet
2. Alcohol
3. Saturation chamber (35 °C)
4. Condensation tube (10 °C)
5. Lamp
6. Photodetector
7. Mass flow meter and pump
8. Slit (0.1 mm × 2 mm)

Figure 3.13. Condensation nucleus counter [68].

lenses collect the light scattered from the particle and focus it on the light detector. The temperature of the optical part is 35 °C, so that the alcohol would not condense on the surfaces of the lenses. The sample aerosol is drawn by a suction pump $(0.3 \, \text{l} \, \text{min}^{-1})$ through an absolute filter and a mass flow meter.

A condensation nucleus counter can operate in two ways. It can count individual particles or the concentration of particles. When the concentration of the aerosol is less than 1000 particles per cubic centimetre, the impulse caused by the light scattered from each individual particle is counted. When the particle concentration is over 1000 particles per cubic centimetre, the light scattered by all those particles is measured which at each moment are simultaneously in the indication space [68].

When counting each particle separately and when calculating the concentration, it is assumed that every particle gives its own light scattering pulse. If there are two or more particles simultaneously in the observation area, they send simultaneously pulses which are calculated as one. The probability of this kind of a coincidence depends on the concentration of aerosol.

The real particle concentration is given by the equation

$$N_a = N_i \exp(N_a q t) \tag{3.18}$$

where N_a is the actual concentration, N_i is the observed concentration, q is the volume flow of the aerosol, and t (e.g. $35 \, \mu s$) is the time which each particle remains in the observation area. The influence of coincidence is shown in Table 3.2.

Table 3.2.   Influence of coincidence on the observation values of the condensation nucleus counter
[69]

Observed concentration (particles cm^{-3})	Correction coefficient	Corrected (particles cm^{-3})
9.00×10^2	1.208	1.088×10^3
8.00×10^2	1.179	9.43×10^2
6.00×10^2	1.125	6.75×10^2
4.00×10^2	1.078	4.31×10^2
1.00×10^2	1.018	1.02×10^2
5.00×10^1	1.009	5.04×10^1
1.00×10^1	1.002	1.00×10^1

Chapter 4

Determination of Metal Emissions

Metal emissions are generated in the combustion of almost all fuels. Table 4.1 represents average metal concentrations for various fuels.

Table 4.1. Average heavy metal concentrations for some fuels. Variation range is given in brackets [1].

Concentration in the fuel (μg/MJ) Element	Heavy fuel oil	Coal	Peat
As	2	100 (20–1000)	100 (20–500)
Cr	1 (1–5)	600 (100–2000)	100 (50–500)
Ni	300 (100–500)	700 (100–3000)	400 (50–1000)
V	1000 (200–2000)	2000 (100–6000)	300 (100–1000)
Pb	20 (2–40)	800 (100–2000)	200 (20–1000)
Cd	0.3	30 (1–200)	10 (5–50)
Hg	0.03	3	5

4.1. Metals Bound by Particles

Emissions of metals bound by particles are generally measured in connection with the determination of the particle concentration [51]. Most commonly, the particles are collected on a quartz or glass fibre filter for the determination of metals. A silver membrane filter can also be used. Samples are collected similarly to when measuring particle emissions, that is to say, by collection inside (in-stack) or outside the channel. In the Nordic countries, an in-stack method is used, according to which the sample is collected by a glass probe in either a quartz fibre filter socket, or in cleaned quartz or glass wool which is on a glass sinter (Fig. 4.1).

Sampling equipment made entirely of glass, designed for the sampling of polycyclic aromatic compounds, has also been used for the collection of metal samples [70]. It is represented schematically in Fig. 4.2. In the equipment, XAD-2 adsorbent has been replaced by impingers for the collection of metals.

In Germany, VDI instructions have been given for the determination of metals from particles of flue gas. According to these instructions, the particles

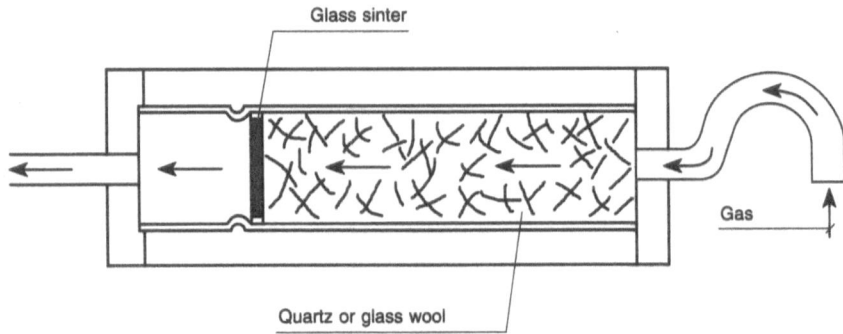

Figure 4.1. Sectional view of the probe used for the collection of metal samples. The interior of the probe is made of glass [51].

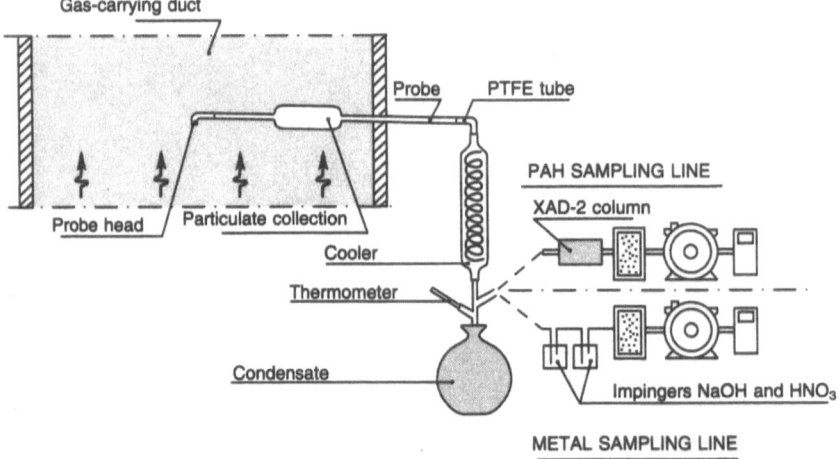

Figure 4.2. Sampling equipment made of glass for the collection of metal emissions [70].

should be collected using in-stack sampling. In the VDI 2268 instructions, detailed methods have been given for the determination of the concentrations of Ba, Be, Cd, Co, Cr, Cu, Ni, Pb, Sr, Th, V and Zn from flue gas particles using an atomic absorption spectrophotometer (AAS) or an optical emission spectrometer (OES). In the method, the samples are dissolved in an open or closed container by treating dust collected in quartz wool with a mixture of strong nitric acid and hydrofluoric acid. The method of sample treatment and analysis according to VDI 2268 is also used in several other countries in Western Europe.

In method 12 of the American EPA system, the particles are collected according to EPA method 5. The collection line corresponding with this

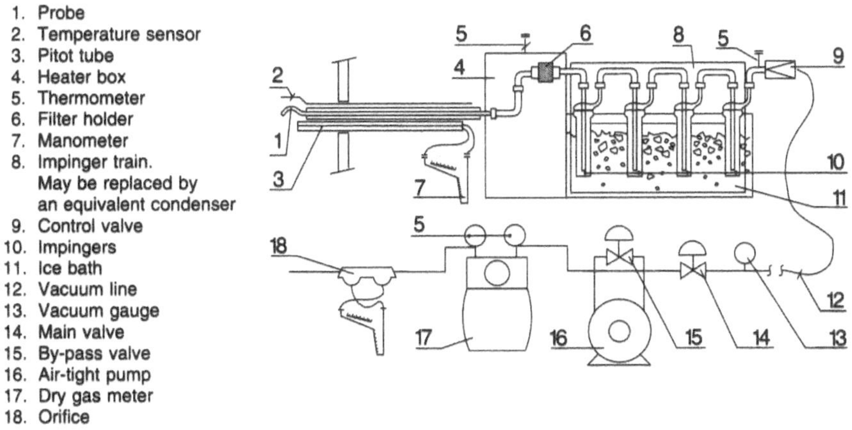

1. Probe
2. Temperature sensor
3. Pitot tube
4. Heater box
5. Thermometer
6. Filter holder
7. Manometer
8. Impinger train.
   May be replaced by
   an equivalent condenser
9. Control valve
10. Impingers
11. Ice bath
12. Vacuum line
13. Vacuum gauge
14. Main valve
15. By-pass valve
16. Air-tight pump
17. Dry gas meter
18. Orifice

Figure 4.3. Collection line corresponding to the EPA method 5 [52].

method is shown in Fig. 4.3. For the determination of lead, 0.1-normal nitric acid is added to the impingers of method 5 for the collection of the lead going through the filter.

The metals can be determined from the particle phase also by particle fractions. This means that the sample has to be collected either with a cascade centripeter, multistage impactor, or cyclone collector.

For the analysis of the metals bound to particles, the following methods can be applied [51]:

- Atomic absorption spectrometry (flame or graphite furnace technique)
- Plasma emission spectrometry
- X-ray spectrometry
- Voltammetry
- Neutron activation analysis
- Proton-induced X-ray emission (PIXE)
- Scanning electron microscopy

The most commonly used methods are atomic absorption, plasma emission and X-ray spectrometry. The samples are decomposed and dissolved for analysis using various kinds of acid mixtures. This stage of the analysis has to be performed carefully so that the substance to be measured is not wasted.

4.2. Metals in Gas Phase

For the purpose of measurement, metals in the gas phase of flue gas refers to metals going through the particle filter [51]. Sometimes the name "gas phase" is in fact replaced by "filter-permeating phase". The metals in the gas phase are collected generally into absorption solutions after the separation of particles.

In the Nordic countries, Germany and Holland, at least the following metals are determined in the gas phase: mercury (Hg), arsenic (As), cadmium (Cd), lead (Pb), copper (Cu), zinc (Zn), magnesium (Mg), chromium (Cr), nickel (Ni), cobalt (Co), and vanadium (V). The sampling time ranges from 15 mins to a few hours.

The metal most commonly determined from gas phase is mercury, which is often found particularly in the gas phase of emission gases. The following absorption solutions are used to collect it [51]:

1. Sequentially two absorption solution bottles, which contain potassium permanganate solution ($KMnO_4$).

2. Sequentially two absorption solution bottles which contain 1% $KMnO_4$ solution and 10% H_2SO_4 solution.

3. Sequentially three absorption bottles; the water-soluble mercury is collected in the first of these, and the mercury non-soluble in water in the two subsequent bottles. There is 10% Na_2CO_3 in one bottle and 1.6% $KMnO_4$ together with 10% H_2SO_4 in two bottles.

4. Four absorption bottles of which the first is empty, and the other three contain $KMnO_4$ and H_2SO_4.

5. Two absorption bottles containing 0.2% $K_2Cr_2O_7$ and 20% HNO_3.

The line used in the sampling of metal emissions is shown in Fig. 4.4. It collects the particle sample isokinetically onto a plane type of quartz fibre

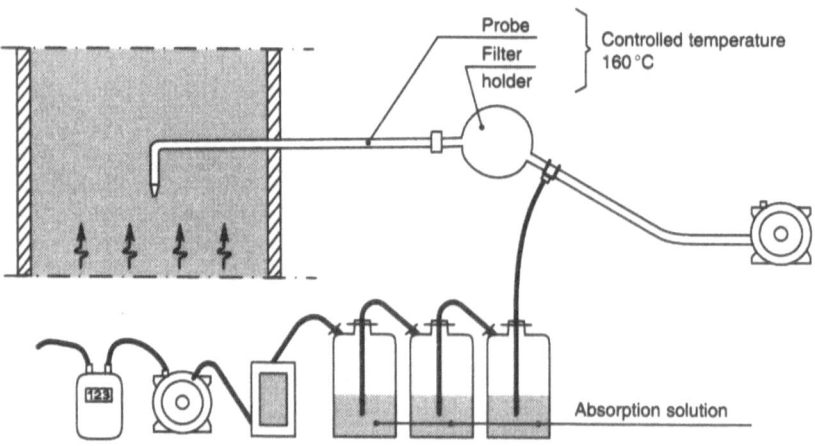

Figure 4.4. Sample line of metal emissions.

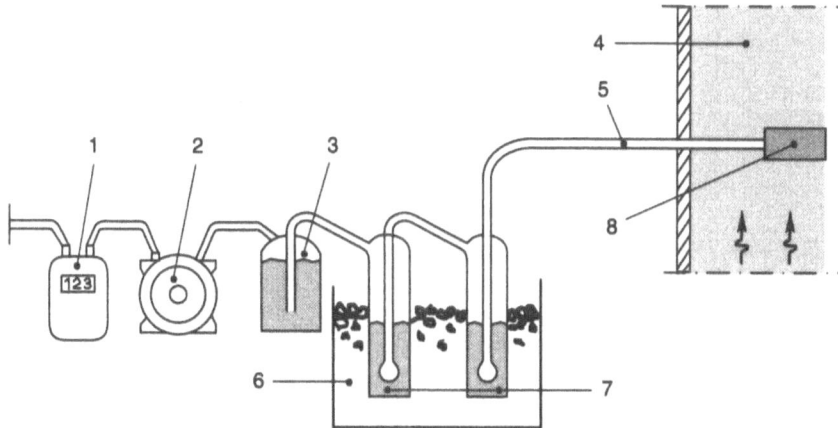

Figure 4.5. Sampling line of gaseous metal emissions. 1, gas volume meter; 2, pump; 3, dryer tube; 4, gas duct; 5, probe; 6, ice bath; 7, impingers; 8, glass wool filter.

filter. The probe and the filter box are kept at a temperature of at least 160 °C. After the filter box, if necessary, one or more secondary lines are taken, the flow rate of which is 2–3 l min^{-1} and which collect the emissions of metals permeating the filter. Through the absorption solution bottles, the gas is directed to a dryer and a volume flow meter. In this method, the absorption solution used is sodium carbonate solution and potassium permanganate–sulphuric acid solution, as listed in (3) above.

Fig. 4.5 represents a typical impinger collection line which is used to collect the metals permeating the filter.

In addition to the methods discussed above, a method modified from the EPA method 5, for example, has been used for the determination of mercury and other metals. It can contain ten impingers.

For the determination of metal emissions, the sample is collected as well as possible, and a quantitative analysis is made, using the methods which are discussed later in the book. This survey concentrates mainly on the sampling of mercury. The sampling of other metals occurring in the gas phase is carried out following the same principle as in the sampling of mercury, but the absorption solution used for the collection is different. Metals other than mercury are most frequently collected in 3% or 10% nitric acid.

A method has also been developed especially for the collection of vaporous metals [51]. Figure 4.6 represents the basic arrangement of collection equipment.

In the collector, the sample gas is mixed with a vaporous, strong acid, and chilled rapidly, so that the vapour is condensed on the surface of the particles and the droplets grow. The grown droplets are collected by means of a filter, cyclone and a collection flask of glass. By means of sampling cyclones used as pre-separators the particles are classified into three size classes.

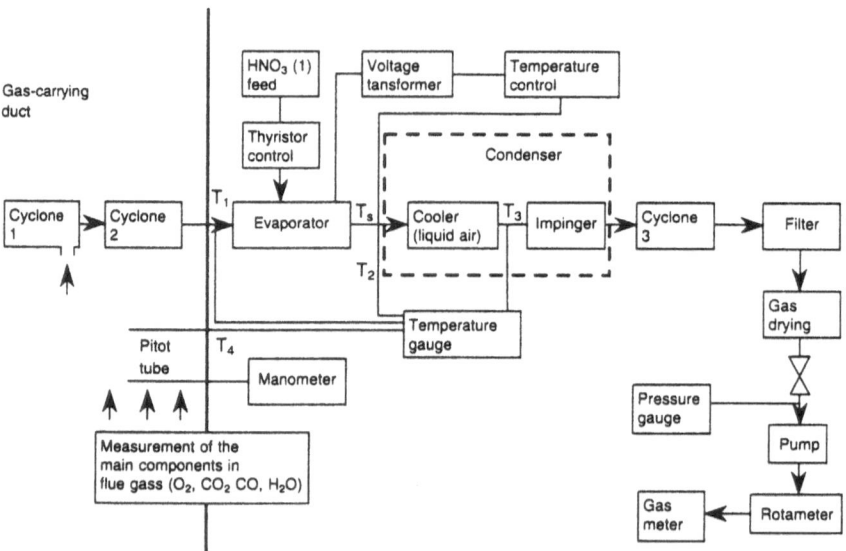

Figure 4.6. Structure of equipment for the collection of heavy metals [71].

Chapter 5

Measurement of Gaseous Emissions

5.1. Sulphur Dioxide

5.1.1. Ultraviolet Fluorescence Method

Molecular Fluorescence

When a gas containing sulphur dioxide is irradiated using ultraviolet light of appropriate wavelength, sulphur dioxide molecules are excited to an energy level higher than the original. When returning to the normal lower energy level, excited molecules give up their extra energy, giving rise to another radiation called fluorescent radiation (also in the ultraviolet wavelength range), the intensity of which is detected. This principle is generally used for the measurement of the concentration of sulphur dioxide in emission gases.

We can compare molecular fluorescence, which we are considering in our discussion of the ultraviolet fluorescence analysis of sulphur dioxide, with atomic fluorescence, which we shall discuss later in connection with X-ray fluorescence. X-ray fluorescence is used for the elemental analysis of solid or liquid samples. Molecular fluorescence is a much more complicated phenomenon than the atomic fluorescence [72].

The atomic energy transitions associated with X-ray fluorescence are much higher than the energy transitions associated with ultraviolet fluorescence, in which molecular energy transitions are involved. This also means shorter wavelengths of the irradiating, as well as the fluorescence, radiations in the case of atomic fluorescence, compared with those in molecular fluorescence. The higher energy needed in the case of atomic fluorescence is due to higher binding energies of the inner shell electrons compared with those of the outer electron energies, as well as vibration and rotation energies in the molecule.

In molecular fluorescence, changes in the energy states of the vibrational and rotational motions are also involved in addition to the electronic transitions. The number of possible transitions thus becomes high and the phenomenon is much more complicated than atomic fluorescence. In our analysis of molecular fluorescence, we deal only with *relaxation processes*, in which molecules return to their basic electronic state from the excited state. This return is a nonradiative process and is called an internal conversion.

Another reason for the complexity of molecular fluorescence is that normally there are other molecules around the excited molecule, so that energy transfer and other relaxation processes with them can take place.

Based on the spin of the electron, either singlet or triplet electronic states can be involved in the excitation process. The excitation is normally taking place from the basic state to a singlet state, as a diamagnetic molecule absorbs light. We pay attention to this type of process. Normally, only two singlet electronic states (S_0 and S_1 in Fig. 5.1 [7.2]) are important in the fluorescence.

In the excitation process, we are also dealing with *absorption*. In excitation, a quantum of energy, e.g. of light, is absorbed by the molecule. This energy is used to raise the molecule to a higher state of energy, i.e. to excite the molecule. This can only be done if there is a match between the energy absorbed and the excitation energy. Excitation can take place between different types of molecular energy, separately or in combination. The molecular energy includes translational, rotational, vibrational and electronic energies. We shall later find the absorption effects important also in connection with infrared techniques used in the measurement of emissions.

Transitions associated with the return of an excited state to ground state are of two basic types: radiative and nonradiative [72]. The molecule is often excited to a higher energy level than the basic excited singlet level (S_1). By a very rapid internal conversion or transfer as thermal energy to the surroundings, the transition occurs to the basic excited singlet level. The

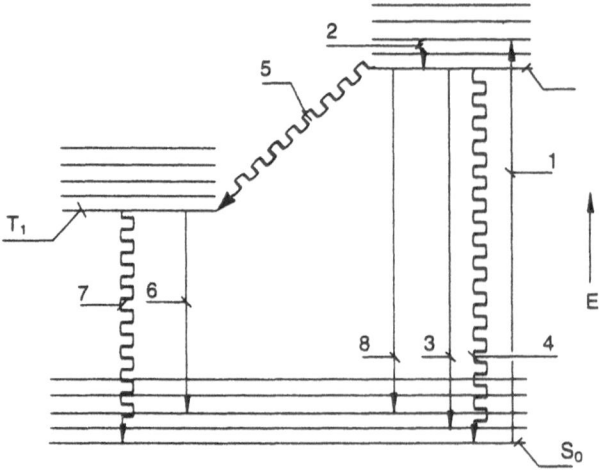

Figure 5.1. Some energy transitions of a molecule. Arrows numbered 3 and 8 indicate fluorescence. The other processes are: 1, absorption; 2, vibration relaxation (nonradiative); 4 and 7, quenching or nonradiative conversion of electronic energy into heat; 5, nonradiative transition between energy systems; 6, phosphorescence. (After H. A. Strobel, *Chemical Instrumentation: a Systematic Approach*, 2nd edn, © 1973 Addison Wesley Publishing Company.)

typical lifetime of the state S_1 is a few nanoseconds. The state S_1 can deactivate to state S_0 generating *fluorescent radiation*. The duration of the fluorescence (after the excitation is stopped) is 1 to 10^3 ns [72]. Another type of radiation is phosphorescence, which is longer-lived than fluorescence and which represents a transition from a triplet state to the ground singlet state.

We can visualise the interrelation between absorption and fluorescence spectra as in Fig. 5.2. As the internal conversions are very fast, fluorescent transitions most often originate from the lowest level in S_1 to different vibrational levels of S_0 [72]. If the vibrational energy levels in the basic and excited electronic energy states are grouped roughly correspondingly, there is also a rough "mirror-image" relationship between these two spectra [74]. The fluorescence spectrum appears at longer wavelengths because of the nonradiative losses in the excited electronic state [72].

In the fluorescent excitation process, the amount of energy needed is the basic electronic energy increase plus the vibrational energy change, whereas in the de-excitation process, the energy released as the fluorescent radiation is the basic electronic excitation energy reduced by the vibrational energy increase (compared with the original situation). This can be concluded from Fig. 5.1 at the points of the arrowheads.

As was noted, the fluorescent excitation normally also yields an increase in the vibrational energy. In some cases, this can result in the dissociation of the molecule. This is normally not very probable, however, as the incident energy flux is small [72]. The vibrational energy should also ordinarily remain high over a sufficiently long period to be redistributed into a vibrational mode, that favours dissociation. It is more probable that vibrational energy will be lost than that dissociation will take place. Molecular dissociation or other photochemical reactions are the more likely, the shorter the wavelength of the irradiating radiation. For this reason, very short wavelengths (e.g. below 200 nm) are avoided.

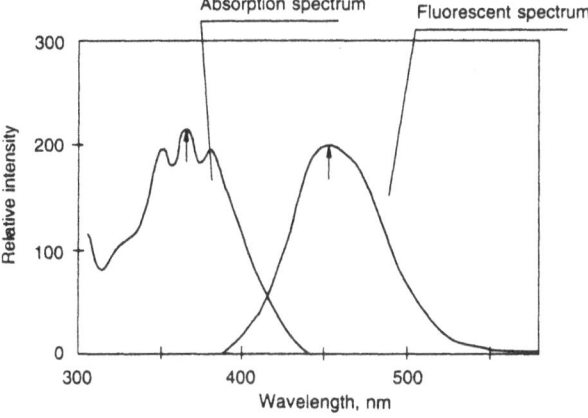

Figure 5.2.   Example of absorption and fluorescent spectra [73].

Nonradiative mechanisms that allow the electronic energy to be degraded to heat also operate [72]. They are called quenching. These include collisions and other mechanisms, e.g. the dissociation of the molecule. The transfer of energy is continued until equilibrium is established.

Wavelength of Irradiating Radiation

As was mentioned earlier, light of a shorter wavelength than 200 nm is not normally used to excite ultraviolet fluorescence in sulphur dioxide. This is because molecular dissociation or other photochemical reactions are the more likely the shorter the wavelength. When measuring sulphur dioxide, the fluorescent yield and the lifetime of the SO_2 fluorescence are strongly reduced under 220 nm due to the dissociation of the excited molecules. As an advantage of the short fluorescent lifetime, quenching by water vapour is minimal. The wavelength of the irradiating radiation used for the measurement of the concentration of sulphur dioxide is thus 214 nm, which is emitted by a zinc lamp. According to a study, water vapour can cause a reduction of up to 5% in the light intensity as a result of the quenching at the relative humidity of 50% [75].

Reactions in Sulphur Dioxide Induced by Ultraviolet Light

In the reaction chamber of the ultraviolet fluorescent analyser the following reactions can take place:

$SO_2 \xrightarrow{I_a} SO_2^*$ \qquad\qquad (creation of an electronically excited state)

$SO_2^* \xrightarrow{k_f} SO_2 + h\nu$ \qquad (de-excitation of the excited state causing fluorescence)

$SO_2^* \xrightarrow{k_d} D$ \qquad\qquad (dissociation)

$SO_2^* + [M] \xrightarrow{k_d} SO_2 + [M]$ \qquad\qquad (quenching)

I_a denotes the intensity absorbed from the excitation radiation (intensity I_0). This absorbed intensity generates fluorescent radiation and other reactions. The intensity of the fluorescent radiation, I_f, is proportional to the intensity absorbed, I_a.

$$I_a = I_0(1 - e^{-\alpha_\lambda c l}) \tag{5.1}$$

where c is the concentration of sulphur dioxide, and $e^{-\alpha_\lambda c l}$ describes the portion of the radiation transmitted through the sample gas according to the Beer–Lambert law.

$$I_f = \frac{k_f I_a}{k_f + k_d + k_q[M]} \tag{5.2}$$

where k_f, k_d and k_q are reaction constants and $[M]$ represents air molecules, which act as quenchers. On the other hand, the intensity absorbed is proportional to the intensity of the irradiating radiation incident on the gas to be measured, I_0. These proportion factors, together with the geometrical factors of the reaction chamber, can be included in one coefficient G, resulting in

$$I_{f,\lambda'} = \frac{G I_0}{k_f + k_d + k_s}(1 - e^{-\alpha_\lambda cl}) \tag{5.3}$$

Here α_λ is the absorption coefficient of sulphur dioxide to the excitation radiation, the wavelength of which is λ, λ' denotes the wavelength of the fluorescent radiation, l is the absorption length and $k_s = k_q[M]$ is a constant associated with the quenching. When the concentration of sulphur dioxide is low and the absorption length short, this expression is reduced to

$$I_{f,\lambda'} = kc \tag{5.4}$$

In practice, a linear relationship is achieved at concentrations lower than 500 ppm [76].

Construction and Operation of Analyser

The construction principle of a sulphur dioxide analyser based on the ultraviolet fluorescence principle is shown schematically in Fig. 5.3. The light source is a zinc glow-discharge lamp, which emits a peak of ultraviolet light at 213.8 nm. Other wavelengths are removed from the beam as far as possible using filters, such as an interference filter or a chlorine filter, which cuts the wavelength band between 270 and 390 nm away almost completely (Fig. 5.4). This filter, however, transmits more than 90% of the light at 214 nm emitted by the zinc lamp [76]. The irradiating light is focused at the centre of the reaction chamber by a lens. A scheme of the structure of an ultraviolet fluorescence sulphur dioxide analyser is represented in Fig. 5.5. The constructional features of the reaction chamber affect the intensity of the fluorescent radiation.

The analyser based on the ultraviolet fluorescence principle measures the concentration of sulphur dioxide continuously. The sample gas is sucked into the reaction chamber at practically the ambient pressure. The fluorescent radiation is detected by a light detector and a photomultiplier tube. The detector unit is normally at right angles to the incident radiation. Another detector and photomultiplier is placed straight across the chamber. This assembly is used for the compensation of the variations in the radiation source. Both the measuring and reference signals are chopped and used for the measuring signal and for the compensation.

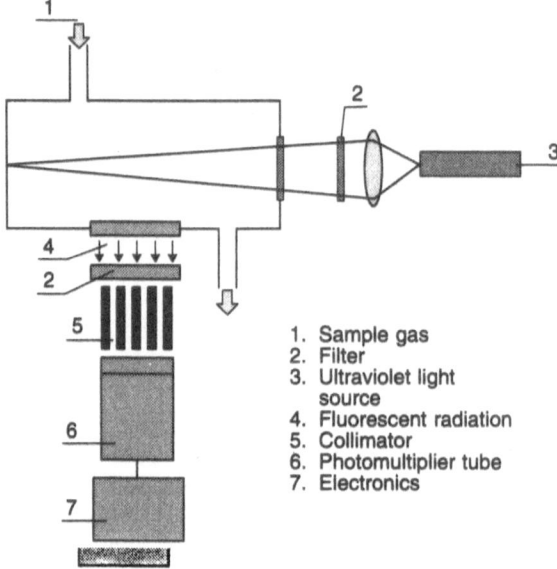

1. Sample gas
2. Filter
3. Ultraviolet light source
4. Fluorescent radiation
5. Collimator
6. Photomultiplier tube
7. Electronics

Figure 5.3. Scheme of the sulphur dioxide analyser based on the ultraviolet fluorescence principle.

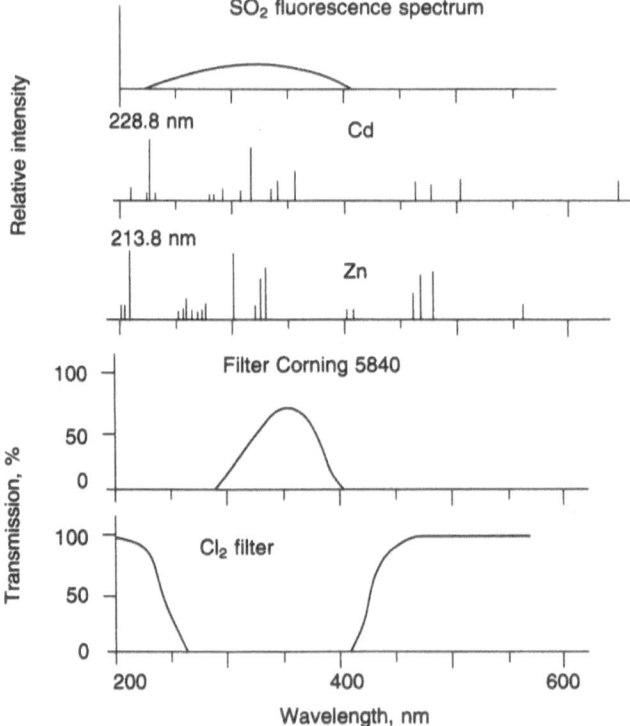

Figure 5.4. Fluorescence spectrum of sulphur dioxide, emission lines of cadmium and zinc lamps and transmission curves for a glass filter and a chlorine filter [76].

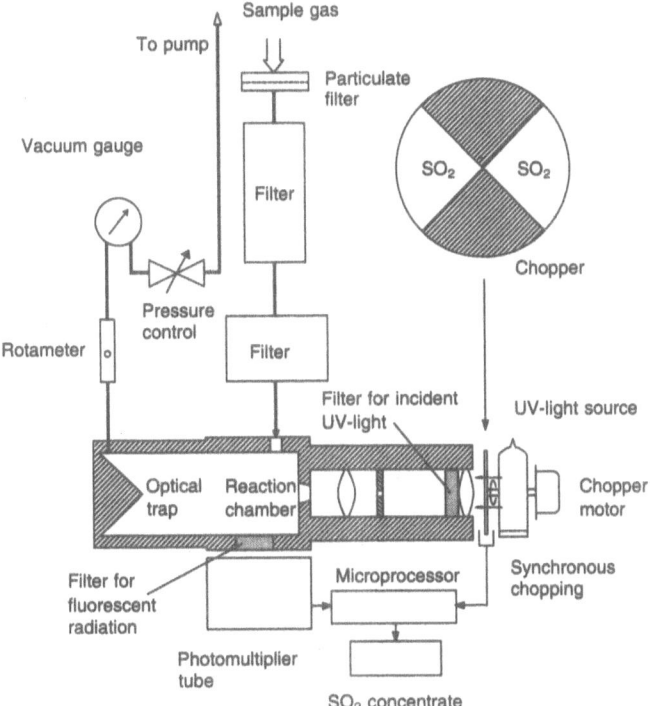

Figure 5.5. Schematic drawing of the structure of an ultraviolet fluorescence sulphur dioxide analyser [77]. Optical trap is used to improve the signal to background ratio.

The temperature affects the amount of gas in the measurement chamber as well as the amplification of the photomultiplier tube. A change of 30°C in the temperature, for example, can change the measurement result about 10%. The reaction chamber and the photomultiplier tube are therefore maintained at a constant temperature.

The sulphur dioxide analyser based on the ultraviolet principle is very sensitive. Its detection limit can be less than 1 ppb (part per billion by volume). When used in emission measurements, the sample gas must be diluted prior to the measurement. This is done using a diluting stack sample.

5.1.2. Method Based on Conductivity of Liquid

In this method, the sulphur dioxide is collected in water to which a small amount of hydrogen peroxide has been added [78]. Sulphur dioxide dissolves in water and becomes rapidly oxidised to sulphuric acid, which dissociates to hydrogen and sulphate ions. These ions increase the electrical conductivity of the solution the greater the amount of sulphur dioxide there is in the gas sucked through the solution. Automatic equipment registers the sulphur dioxide concentration as hourly, daily or monthly means. The method is used primarily in measurements of air quality to register the sulphur dioxide concentration.

5.1.3. Features of Different Methods

The ultraviolet fluorescent method is extensively used for the continuous measurement of the concentration of sulphur dioxide from emission gases. In emission measurements, it is usually operated via a diluting stack sampler. This arrangement overcomes the possible disturbances caused by condensing water vapour, as the moisture level of the gas led into the analyser is reduced to a fraction of the original by the dry diluting air. The concentration of sulphur dioxide is still determined from the actual emission gas without the removal of water vapour.

Without the dilution unit, the ultraviolet fluorescence method is sensitive enough to be used for the air quality mesurements. It is also generally used for that purpose.

Other possible methods for the continuous monitoring of the concentration of sulphur dioxide in emission gases include solid state sensors. These sensors are still under development, but appear promising. These sensors can be placed in emission gas ducts thus allowing continuous monitoring. They are even faster than analysers, as they do not need a sample line outside the duct. For process control purposes, for instance, high temperature solid state sensors offer a promising alternative.

The recent development of Fourier transform infrared (FTIR) analysers has created instruments that can be applied to the measurement of emissions at their sources [79]. Both constructional features and mathematical operations have been developed towards better usability and field operation. These instruments operate almost continuously and can record simultaneously the concentrations of several gas components from emission gases, including sulphur dioxide. FTIR analysers are versatile instruments and are increasingly applied to practical process emission measurements.

5.2. Hydrogen Sulphide and Other Reduced Sulphur Compounds

To measure both hydrogen sulphide and reduced organic sulphur compounds, a technique can be used in which these reduced sulphur compounds are oxidised thermally to sulphur dioxide and measured thereafter by using methods applicable to the measurement of sulphur dioxide concentrations. One method is, for instance, the technique based on ultraviolet fluorescence.

By this method it is possible to determine the total concentration of reduced sulphur compounds or what is called the concentration of the total reduced sulphur (TRS) compounds, as for example in pulp plants. The temperature of the oxidation furnace is about 800 °C. The flue gas must contain a minimum of 1% oxygen in order that all TRS compounds would become fully oxidised to sulphur dioxide.

If the emission gas to be measured contains sulphur dioxide, this has to be scrubbed away from the gas before the oxidation of the reduced compounds. The emission gas is scrubbed using an SO_2 scrubber. This may contain citrate buffer solution (potassium citrate or sodium citrate). The collection efficiency of the sulphur dioxide has to be at least 99%.

Another method for the determination of TRS compounds is titration with barium perchlorate using the thorini indicator. For sampling, the volume of the gas used has naturally to be determined.

5.3. Nitrogen Oxides

The most commonly used method for the measurement of the concentration of nitrogen oxides is based on chemiluminescence, that is to say, generation of light as a result of a chemical reaction. This takes place when nitrogen monoxide and ozone react with each other. The measurement principle can also be applied to the measurement of concentrations of other nitrogen oxides by converting them first into nitrogen monoxide in a catalytic converter. This method is also very sensitive. It can be used for air quality measurements, and its detection limit is in the range of a few ppb (parts per billion by volume).

5.3.1. Chemiluminescence

Some of the reaction products developed in a few chemical reactions remain in excited state, and radiate light when the excitation is discharged. This phenomenon is called chemiluminescence. Particularly at low pressures, at which the collision frequency is low, the excitation is discharged as radiation of light. Chemiluminescence is an opposite reaction to a chemical reaction induced by light. In addition to the chemiluminescence, the extra energy bound to the excited molecule can discharge through impacts or the dissociation of the molecule.

The chemiluminescence reaction between nitrogen monoxide and ozone can be formulated as follows:

$$NO + O_3 \longrightarrow NO_2^* + O_2$$

$$NO_2^* \longrightarrow NO_2 + h\nu$$

Here NO_2^* refers to the excited nitrogen oxide molecule. Of the nitrogen dioxide generated as a result of the reaction above, approximately one tenth is estimated to be in an excited state (at room temperature) [80].

The excitation generated by the chemical reaction can, in addition to radiation, be discharged by quenching or through another chemical reaction:

$$NO_2^* + NO_2 \xrightarrow{k_{q1}} 2NO_2$$

$$NO_2^* + M \xrightarrow{k_{q2}} NO_2 + M$$

$$NO_2^* + NO_2 \xrightarrow{k} 2NO + O_2$$

The quenching can take place by the action of another nitrogen oxide molecule or other molecule, as shown in the first and second reaction equations. The third reaction equation represents the discharging of excitation through a chemical reaction [81].

An instrument for measuring nitrogen oxides based on chemiluminescence is shown in Fig. 5.6. The ozone needed for the reaction is produced in the ozone generator, which is part of the device. One of the reaction chamber walls is an optical filter through which a photomultiplier tube measures the intensity of the chemiluminescence radiation and converts it to a current signal [83]. When the active NO_2 molecules return to their normal state, broadband light is generated, the wavelength of which varies between 500 and 3000 nm and the intensity of which is at its maximum at 1100 nm.

The method is very sensitive and is capable of measuring concentrations of a few ppb by volume. In the emission measurements, the gas sample is conducted into the chemiluminescence analyser through a diluting stack sampler.

The above discussion reveals that the device for nitrogen oxides based on chemiluminescence measures the concentration of nitrogen monoxide. By using

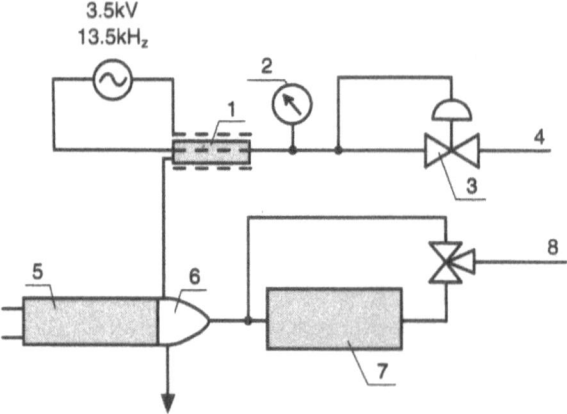

Figure 5.6. Schematic diagram of an instrument for the measurement of nitrogen oxides based on chemiluminescence [82]. 1, ozone generator; 2, pressure gauge; 3, oxygen control; 4, air; 5, photomultiplier tube; 6, reaction chamber; 7, NO_x converter; 8, sample gas.

the same equipment it is, however, also possible to measure the concentration of nitrogen dioxide. Nitrogen dioxide is reduced to nitrogen monoxide in a converter by means of a molybdenum catalyst. In order to get a reliable measurement result, the converter must be efficient (conversion efficiency over 95%).

In instruments measuring nitrogen oxides, the sample gas is often divided into two lines. Of these, one leads directly to the measurement chamber, and it is used to measure the concentration of nitrogen monoxide. The other line runs through the converter, which results in the measurement of the total concentration of nitrogen monoxide and other nitrogen oxides. The majority of the other nitrogen oxides is nitrogen dioxide. The difference of the measurement results of these two channels is the concentration of nitrogen dioxide.

5.4. Hydrocarbons

The concentration of hydrocarbons can be measured by means of the FTIR technique, which will be discussed later. This is a versatile and effective analysing technique for the measurement of concentrations of very many gases. For the measurement of hydrocarbon concentrations, however, instruments based on the use of a flame ionisation detector technology have been developed. In addition to hydrocarbon analysers, flame ionisation detectors are utilised as detectors of gas chromatographs, as revealed further below.

In a hydrocarbon analyser based on flame ionisation, the sample gas is conducted along a heated sampling line to the detector, in the hydrogen flame of which the hydrocarbons are ionised into electrons and positive ions.

The operating principle of the flame ionisation detector is represented in Fig. 5.7. The detector consists of a combustion chamber and a burner. Pure hydrogen is conducted through the burner nozzle and the combustion air through a hole around the nozzle into the combustion chamber. The ions and electrons are collected using the collector electrode placed near the flame. The combustion nozzle can serve as one of the electrodes, and the current flowing between the electrodes can be registered as the signal indicating the concentration.

The flame of pure hydrogen produces a very low concentration of ions. Instead, compounds brought into the flame, which contain bonds of carbon and hydrogen, produce carbon ions in the flame [84]. By using a flame ionisation detector most compounds can be measured which have a bond of carbon and hydrogen. When the gas to be measured contains several hydrocarbons, the resulting response depends on the carbon number. For instance, 10 ppm of methane (CH_4) and 10 ppm of propane (C_3H_8) in air produce a measurement result which corresponds to a concentration of 40 ppm of methane.

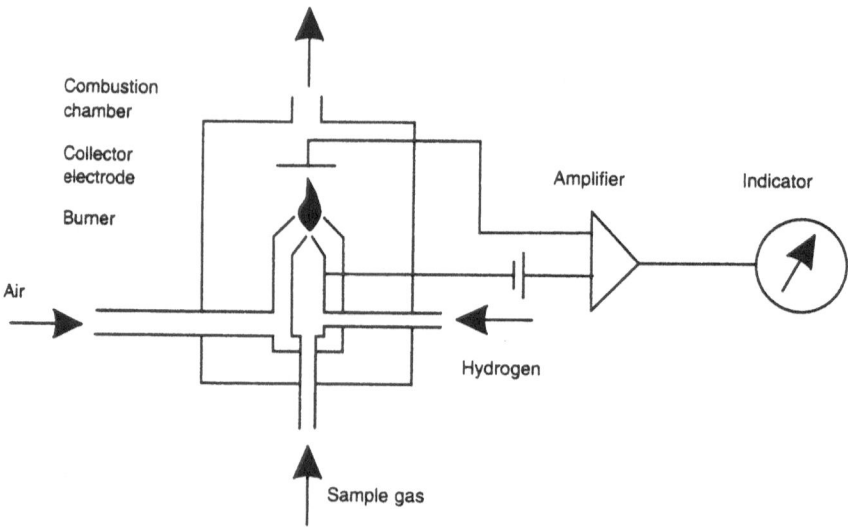

Figure 5.7.   Structural diagram of a flame ionisation detector [51].

The flame ionisation detector is capable of measuring only gaseous hydrocarbons, in other words hydrocarbons which have a low boiling temperature. Emission gases can, however, also contain hydrocarbons which are in liquid form at ambient temperature and pressure. Therefore analysers based on flame ionisation detection are generally equipped with heating elements which keep the sampling line and the detector at a temperature of about 200°C.

Variation in the concentration of oxygen may affect the response of the flame ionisation analyser. The concentration of oxygen can vary considerably, for instance, when measuring hydrocarbon concentrations of emission gases from combustion processes. To minimise the error caused by this effect, the fuel used can be a mixture of hydrogen and helium, of which hydrogen accounts for 40% and helium 60%. The flow rate of the fuel has to be increased then so that the flow of hydrogen remains the same as without helium [84].

A measuring device based on the flame ionisation detector measures, consequently, the total concentration of hydrocarbons. By using a catalyst, for example a heated platinum wire, hydrocarbons other than methane can be removed from the sample gas. This means that the total concentration of hydrocarbons, the concentration of methane, and the concentration of hydrocarbons other than methane can be determined.

5.5. Carbon Monoxide and Carbon Dioxide

When measuring the concentrations of carbon monoxide and carbon dioxide from emission gases, the equipment most frequently used is based on the absorption of infrared light. The operation of these devices is discussed below. In this connection, the example examined is the operation principle of a type of gas analyser [85–87].

The infrared analysers used for the measurement of the concentrations of carbon monoxide and carbon dioxide generally utilise the characteristic wavelength of infrared light. The gas to be measured absorbs this wavelength strongly. The wavelength band is usually selected using an optical bandpass filter. This kind of equipment is called a nondispersive infrared (NDIR) gas analyser. The term is used to distinguish this type of measuring device from dispersive gas analysers, which are generally based on the analysis of a broader wavelength band. These include the Fourier transform infrared (FTIR) analyser, to be discussed below.

Figure 5.8 shows the operation principle and an arrangement of a nondispersive gas analyser. The analyser consists of an optical bench, light source, detector and measurement electronics.

The optical arrangement of the apparatus comprises an infrared light source, a 50 Hz beam chopper, sampling and reference chambers, as well as the detector part where the optical filter required for the gas to be measured is situated. This lets through only that infrared wavelength band which the gas to be measured will absorb. The infrared light source is a glowing filament wire or metal bar whose temperature is over 800°C. The radiation source most commonly used is tungsten wire or chromium nickel wire.

The infrared light of the source is focused into two beams by means of a concave mirror. One of these passes through the sample chamber and the other through the reference chamber. By using the chopper, the light is let in turn through both chambers. As the light beam passes though the sample chamber, the gas within it selectively absorbs light, since the wavelength has been chosen at the peak absorption. Selective absorption does not, on the other hand, take place in the reference chamber, which has been filled with air or nitrogen. The detector, e.g. a semiconductor sensor, in turn receives an infrared beam which has passed through the sample chamber or the reference chamber. By comparing the responses from these in the detector, the magnitude of the selective absorption caused by the sample gas can be determined and the concentration of the gas measured can be calculated. The beam passing through the reference chamber is utilised to eliminate errors caused by possible variations in the infrared light source or the detector sensitivity.

5.6. Oxygen

The oxygen concentration of emission gases has to be known to be able to follow the process and to calculate the specific emission, for example the emission in a combustion process per fuel energy unit (e.g. $mg\,MJ^{-1}$). For the continuous measurement of the oxygen concentration from emission gases, a

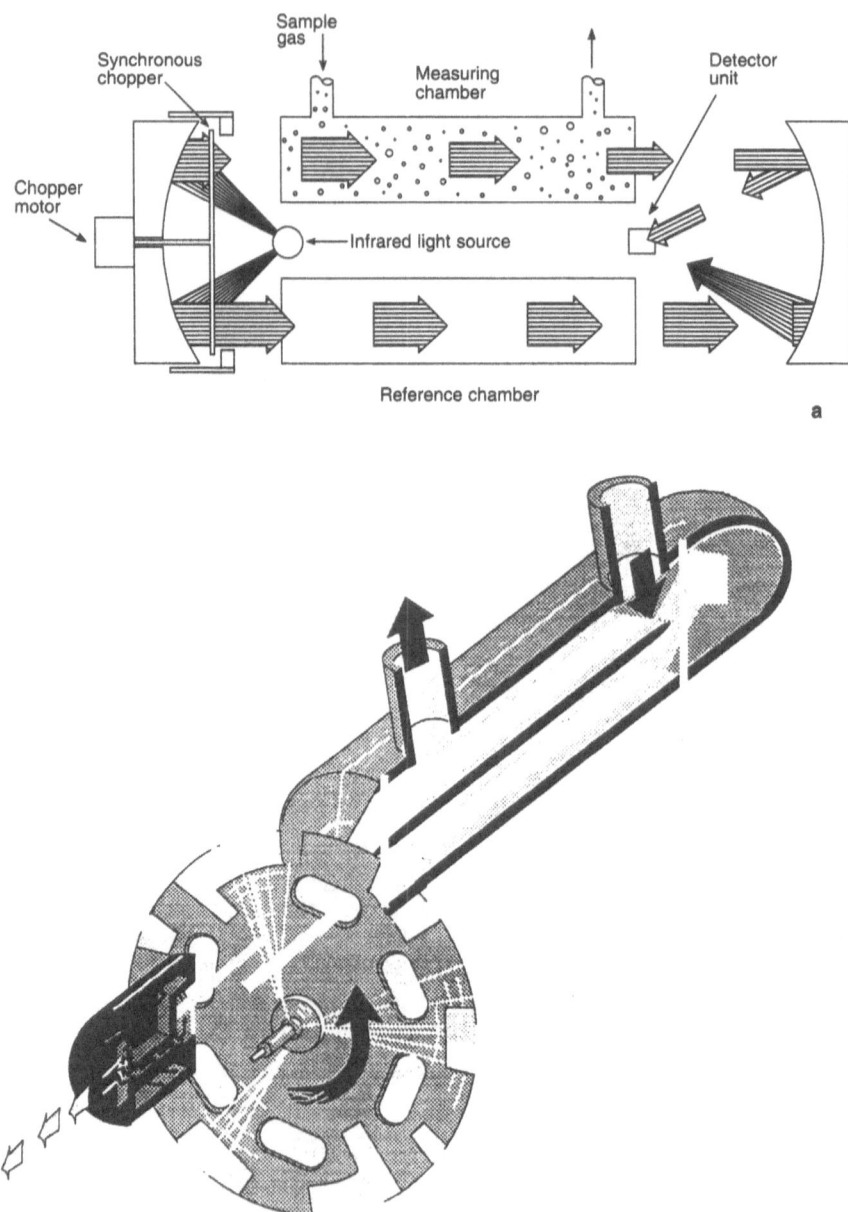

Figure 5.8. **a** Operation principle [85] and **b** arrangement [87] of a nondispersive infrared analyser.

commonly used method is based on the paramagnetism of oxygen. Another method is based on a zirconia cell, between the different sides of which an electrical voltage is generated as a function of the ratio of the partial pressures at both sides. The paramagnetic instrument requires a sampling line to conduct the sample gas to the measuring device, whereas the zirconia cell can be placed directly in the emission channel. In the following, operation principles of paramagnetic instruments for the measurement of the concentration of oxygen are represented. The principles of the operation of the zirconia cell are considered in connection with the discussion of sensor technologies for the measurement of emission gases.

5.6.1. Paramagnetism

All substances have magnetic properties. They are, first of all, due to the fact that in atoms the orbit of each electron forms a closed current loop, which has a magnetic moment. On the other hand, each electron spins on its own axis, and this spinning charge contributes to the magnetic moment. The proportions of these two factors, that is to say, orbit and spin components, have to be known in order that the magnetic moment can be calculated.

The movement of the electron around the nucleus is characterised by four quantum numbers. The main quantum number n determines principally the energy [88,89]. The angular momentum quantum number l involves a magnetic orbital dipole moment, and this dipole can orient itself in a magnetic field only to definite directions in relation to the field. The energy of the dipole depends on its orientation, and this is determined by the values of the magnetic quantum number m. All the aforementioned quantum numbers are integers. Additionally, the electron has a spin quantum number s whose values may be $\pm\frac{1}{2}$.

The electrons fill the lowest energy states possible, following the Pauli exclusion principle. Thus only one electron can be in a state characterised by certain values of n, l, m and s. If all states corresponding to the quantum numbers n and l are full, their total spin dipole moment is zero, because the number of electrons whose $s = \frac{1}{2}$ equals the number of electrons whose $s = -\frac{1}{2}$. Their orbital dipole moment in the presence of a field is also zero, because there is an electron in state $-m$ for each electron in state $+m$. This kind of electron group forms a closed shell.

Consequently, only electrons in incomplete shells can affect the magnetic dipole moment. Other magnetic phenomena except diamagnetism occur, based on the above, only in such substances in which there is an incomplete shell in addition to that formed by the valence electrons.

In addition to diamagnetism, which is found in all substances, magnetic phenomena include paramagnetism and ferromagnetism (others are ferri- and antiferromagnetism). Of these, dia- and paramagnetism represent field magnetism, and a magnetic dipole moment can develop in them only by the action of an external magnetic field. Ferro-, antiferro- and ferrimagnetic substances can have a magnetic dipole moment even without an external magnetic field. These types of magnetism represent order magnetism.

Paramagnetic substances move in the magnetic field, as much as possible, towards the strongest part of the field, whereas the force acting on diamagnetic substances tends to move them away from the strongest part of the field.

In the molecules of most gases there is an even number of electrons, which means that they are diamagnets. Instead, in the molecules of NO, NO_2 and CO_2, for instance, there is an odd number of electrons, and they are naturally paramagnetic [90]. Oxygen is a special case, that is, it does have an even number of electrons, but when they are assigned to their molecular orbitals, two are left with spins unpaired. Oxygen is therefore paramagnetic.

5.6.2. Magnetic Susceptibility

The strength of the magnetic behaviour of a substance is characterised by the magnetic susceptibility κ_V, which is the ratio of the magnetic moment (m_V) to the magnetic field intensity H, or

$$\kappa_V = \frac{m_V}{H} \tag{5.5}$$

This refers to quantities calculated per volume unit, which is symbolised by the subscript V. The susceptibility of ferro- and paramagnetic substances has a positive sign, whereas the susceptibility of diamagnetic substances is negative. Table 5.1 shows relative susceptibilities for some gases. The susceptibility of oxygen is denoted by 100. According to Curie's law, the mass susceptibility of a substance is inversely proportional to the absolute temperature. Because the

Table 5.1. Magnetic susceptibilities for certain gases relative to oxygen (100) and nitrogen (0) [90]. Of the gases shown, only O_2, NO and NO_2 are paramagnetic. The + sign in connection with other gases is caused by the rearrangement of the scale so that the susceptibility of nitrogen is zero

Acetylene	−0.24	n-Hexane	−1.7
Ammonia	−0.26	Hydrogen	+0.24
Argon	−0.22	Hydrogen chloride	−0.30
Bromine	−1.3	Hydrogen fluoride	+0.10
Butadiene 1:2	−0.65	Hydrogen sulphide	−0.39
Butadiene 1:3	−0.49	Krypton	−0.51
n-Butane	−1.3	Methane	−0.2
i-Butane	−1.3	Neon	+0.13
Butene-1	−0.85	Nitric oxide (NO)	+43
i-Butene-1	−0.89	Nitrogen dioxide	+28
c-Butene-2	−0.85	Nitrogen	0
t-Butene-2	−0.92	n-Octane	−2.5
Carbon dioxide	−0.27	Oxygen	100
Carbon monoxide	+0.01	n-Pentane	−1.45
Cyclohexane	−1.557	i-Pentane	−1.49
Ethane	−0.46	Propane	−0.86
Ethylene	−0.26	Propylene	−0.545
Helium	+0.30	Water	−0.02
n-Heptane	−2.1	Xenon	−0.95

volume susceptibility is the mass susceptibility multiplied by the density, and because density is also inversely proportional to the absolute temperature, the volume susceptibility is inversely proportional to the second power of the absolute temperature, that is to say,

$$\kappa_V \propto \frac{1}{T^2} \tag{5.6}$$

This proportionality forms the basis for the thermomagnetic measurement of the concentration of oxygen. On this basis it can further be noted that the measurement of the concentration of oxygen based on susceptibility requires precise control of the temperature.

5.6.3. Method Based on Paramagnetic Principle

The concentration of oxygen can be determined by measuring the magnetic susceptibility directly. More often, a heater and a resistor balance bridge are utilised in the paramagnetic oxygen measurement. These kinds of measuring devices are known as thermomagnetic measuring instruments, and which utilise "the magnetic wind". In all types of equipment, permanent magnets are used.

Figure 5.9 shows the principle of a thermomagnetic instrument for the measurement of the concentration of oxygen.

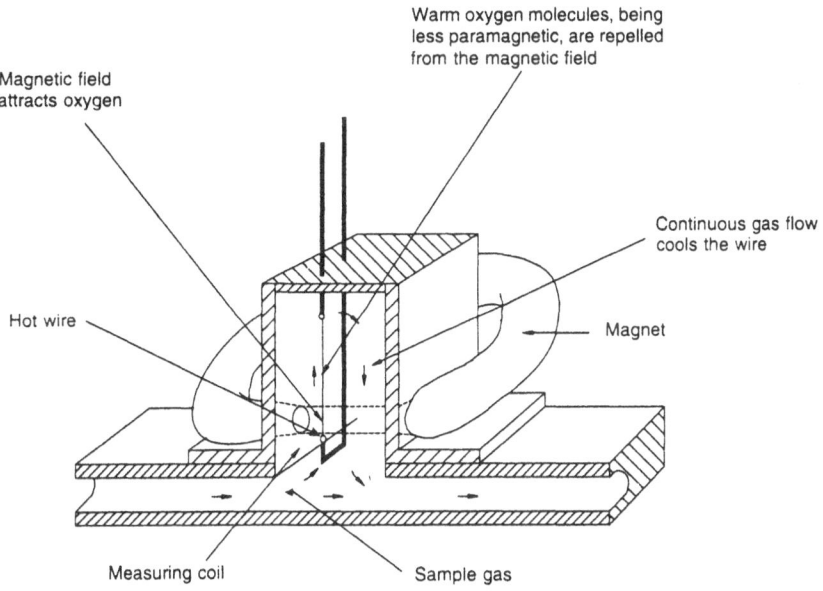

Figure 5.9. Paramagnetic oxygen cell based on the measurement of heat conduction [30].

In a thermomagnetic meter, the gas flows through a measuring cell. This cell contains a heating element in a strong magnetic field. When the sample gas contains oxygen, the magnetic field attracts it. When the heating element heats the gas this becomes, as stated above, less magnetic and is displaced to make way for the colder and more intensely magnetic gas. This leads to a flow which cools the heating element, altering its resistance simultaneously. Based on this, the concentration of oxygen can be determined. In this arrangement, the flow rate depends primarily on the temperature and on the magnetic field intensity as well as on the magnetic susceptibility of the gas. The temperature and the magnetic field intensity are constant and the magnetic susceptibility of the gas is proportional to its oxygen concentration. Other properties of the gas naturally also affect the heat conduction, and thus the resistance of the heating element, but their effects can generally be diminished to insignificance.

In a ring chamber analyser, the sample gas flows along both sides of the ring-shaped sample cell (Fig. 5.10). In the middle of this ring there is a horizontal glass tube, which connects different sides of the ring cell. This glass tube is surrounded by two similar resistor coils, one of which is in the magnetic field. If the sample gas contains oxygen, the magnetic field attracts it to the connecting tube where the gas becomes warmer. The magnetic susceptibility, on the other hand, is decreased by heat, and the magnetic field attracts cold gas more strongly. A flow is developed, which transports heat from the section in the

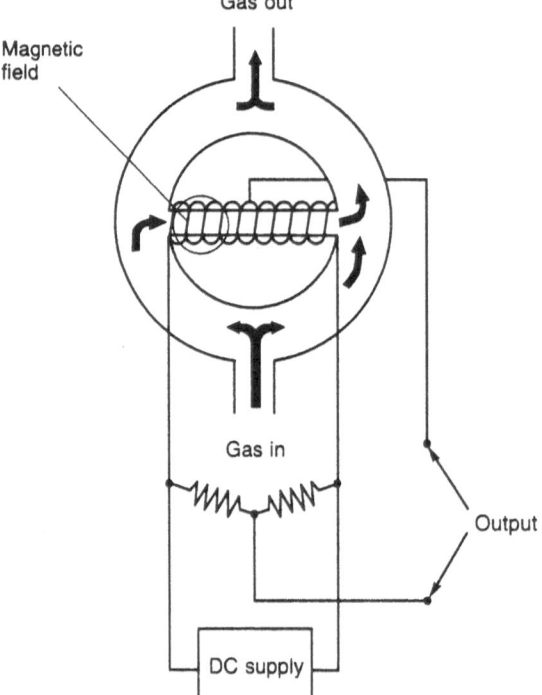

Figure 5.10. Ring chamber of a paramagnetic oxygen analyser (Kent) [90].

magnetic field to the other end of the connecting tube (in the figure from left to right). This means that the temperatures of the resistor coils become unequal and cause the resistance bridge to become unbalanced. The measurement result it gives then depends on the oxygen concentration of the sample gas.

It should be noted that under usual measuring conditions the concentrations of other paramagnetic gases, such as nitrogen monoxide and nitrogen dioxide are very small compared with the concentration of oxygen. Consequently, they do not cause any significant errors. For example in a flue gas, the volumetric ratio of oxygen and nitrogen monoxide can be of the order of 100, and as the magnetic susceptibility of nitrogen monoxide is lower than half of the value for oxygen, the error caused by it remains smaller than 0.5%.

The measurement of the concentration of oxygen using a zirconia cell is discussed later.

5.7. Diluting Stack Sampler in Emission Measurements

To transfer the gas sample from the sampling site to the gas analyser, a diluting stack sampler or a dilution unit is often used, especially with the ultraviolet fluorescence SO_2 analyser and the chemiluminescence NO_x analyser. The system dilutes the gas sample so that only $50-100\,\mathrm{ml\,min^{-1}}$, that is, approximately 5% of the total flow normally required by the analyser, is taken [91]. The diluting air used is dry instrument air the pressure of which is about 5 bar, the volumetric flow being about $5\,\mathrm{l\,min^{-1}}$. In the probe belonging to the system the diluting air is warmed to the same temperature with the sample gas, which means that the dilution ratio remains constant.

The flow of the diluting air causes an ejection effect which aspirates the sample gas through a very thin glass tube or the critical orifice (Fig. 5.11). At the same time the sample gas and the dilution air are mixed.

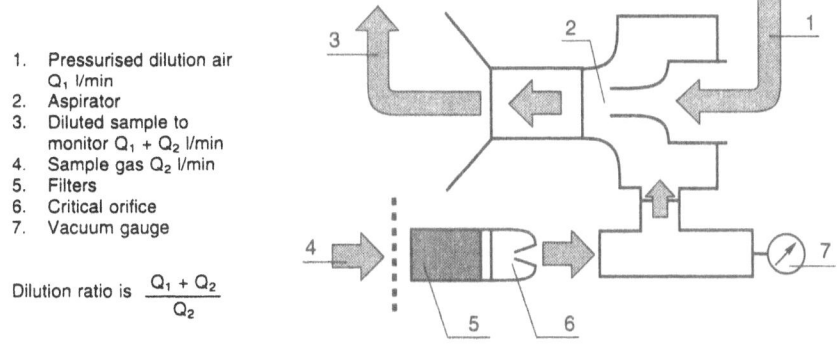

1. Pressurised dilution air Q_1 l/min
2. Aspirator
3. Diluted sample to monitor $Q_1 + Q_2$ l/min
4. Sample gas Q_2 l/min
5. Filters
6. Critical orifice
7. Vacuum gauge

Dilution ratio is $\dfrac{Q_1 + Q_2}{Q_2}$

Figure 5.11. Operation principle of the diluting stack sampler [91].

From the control unit of the diluting sampler system the flows are arranged, with which sample gases are taken to the analyser, their concentration being 10–50 ppm. The optimal measuring ranges of some methods, such as the NO_x measurement based on chemiluminescence and the SO_2 measurement based on ultraviolet fluorescence, fall upon these concentrations when their original concentrations in emission gases are of the order of hundreds of ppm. The control unit is generally located near the analyser.

Because the dilution air (which consequently forms the major part of the gas going into the analyser) used by the dilution probe is dry, no cooling or heated sampling line is needed for the transfer of the sample. Due to the small flow of the sample gas, only a small amount of impurities reaches the analyser.

The dilution unit is calibrated by conducting the calibration gas along a flexible tube line to the sampling site. The calibration thus concerns the whole system. Its accuracy and reliability depend almost entirely on whether the concentration of the control gas holds true [91].

5.8. Determination of Organic Compounds

5.8.1. Sampling

For the sampling of polycyclic and chlorinated aromatic compounds from the flue gas, a method has to be employed which yields a sample of both the particle and gas phase [10]. The basic components of the sampling equipment are sampling probe, filter, cooler, adsorbent or absorbent, pump and control equipment.

Two versions of the method developed by EPA for the sampling of flue gases are in use: the basic model whose suction rate is about $1.5\,m^3\,h^{-1}$, and the power collector model whose suction rate is about $8\,m^3h^{-1}$ [92]. The assembly is shown in Fig. 5.12. The main components are sampling probe, particle separator, cooler, adsorbent and absorbent units, as well as control equipment. Those parts of the apparatus which come into contact with the flue gas to be analysed are made either of glass or polytetrafluoroethylene (PTFE).

The sampling probe is all glass, and it is equipped with a heating resistor. For the separation of particles, either a cyclone and a filter, or just a filter, is used. The cyclone can be used to separate the majority of the particles and thus to prevent blocking of the filter. The particles are separated in a box fitted with a thermostat, the temperature of which during sampling is $120°C \pm 12°C$. After filtering, the flue gas is cooled to the temperature of $+20°C$. The cooling and adsorbent unit recommended by EPA is like the one shown in Fig. 5.13. Normally, there is only one resin layer (adsorbent) in the basic model. In the power collector version, two resin layers are used and the condensate is separated before the gas is conducted through the resin.

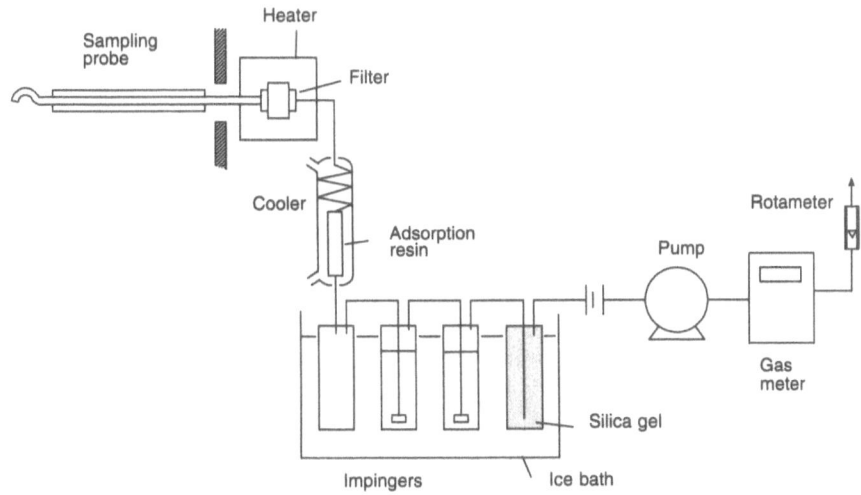

Figure 5.12. Basic model of the sampling apparatus of the EPA Modified Method 5 (MM5) [93].

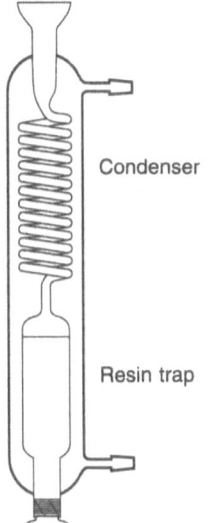

Figure 5.13. Adsorption unit of the EPA basic model [93].

When using the basic method MM5, the sample is taken, as far as possible, isokinetically. A typical sampling time is 3–5 h, in which case the total volume of the gas collected is 5–8 m^3. The sampling method is suitable for the collection of PCB and PAH compounds as well as chlorinated organic compounds having the carbon atom number 7–16 and the chlorine atom number 1–10 [92]. The applicability of the equipment of the EPA methods to field use is restricted by the fact that they have a great number of glass components.

The source assessment sampling system (SASS) is an apparatus for sampling of organic compounds, designed for collection of high volume samples. During a sampling period of 3 h, the apparatus collects approximately 30 m^3 of sample. As far as the principle is concerned, the method is the same as MM5.

The sampling probe of the SASS equipment is of stainless steel, and part of the solid matter is separated from the flue gas by means of cyclones before it goes to the filter. Usually, the set has three cyclones in series, which classify the particles into size classes $> 10\ \mu m$, $10–3\ \mu m$, and $3–1\ \mu m$. Both the cyclones and the filter are fitted with thermostats. After the separation of the particles, the flue gas is cooled and led through the adsorbent. The adsorbent used is XAD-2 resin.

The advantages of this sampling method are the high sample volume and the possibility for the fractioning of particles. However, acid gases may cause corrosion to the apparatus; also the apparatus is heavy and clumsy.

The basic composition of the equipment used for the sampling of organic chlorine compounds and PAH compounds in the Nordic countries is shown in Fig. 5.14.

The components of the equipment which get into contact with the sample are of glass or PTFE. The temperature of the thermostat controlled filter is usually 120–160°C during the sampling. After the filtering, the flue gas is cooled to a temperature below 20°C and the condensed moisture is separated before conducting the gas through the adsorbent. The flue gas sample is taken isokinetically and the usual sampling time is 3–4 h. The volume of the sample collected is then about 10–20 m^3.

In Germany, a method based on filtering and liquid absorption is recommended for PCDD and PCDF samples. In this method, the temperature of the filter is adjusted during the sampling to correspond to the temperature of the sampling site in the flue gas duct. After the filtering, the flue gas is cooled and the condensed water separated. The organic gaseous compounds are absorbed in bubbling collectors into ethylene glycol monomethyl ether, or

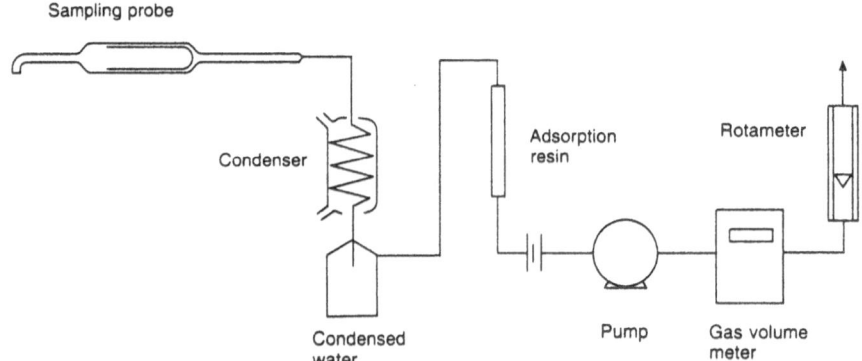

Figure 5.14. Sampling equipment for organic chlorine compounds and PAH compounds [10].

ethoxy ethanol [94]. The volumetric flow of sampling is about $4\,m^3\,h^{-1}$ and the volume of the samples collected is about $20\,m^3$.

5.8.2. Adsorption Materials

Solid adsorbent is most commonly used for the sampling of gaseous organic compounds [10]. The most important properties of the sampling material are heat resistance, particle size and size distribution, pore size and size distribution, surface area and chemical properties. These include the resistance to acids, solvents and reactive gases, as well as the polarity and the background concentrations. These properties influence, for example, pressure losses, resistance to flue gas conditions, adsorption behaviour and yield of the compounds to be studied. Table 5.2 shows adsorption materials used for the sampling of the most common organic compounds. Tenax and XAD resins are generally used for both flue gas and outdoor sampling. Polyurethane foam has mostly been used for the collection of outdoor samples.

Table 5.2. Properties of adsorption materials [95, 96]

Adsorption material	Chemical structure	Specific surface area $m^2\,g^{-1}$	Pore size (nm)	Maximum temperature
Tenax GC	Diphenyl phenylene oxide	20–25	72	375
XAD-2	Styrene divinyl benzene	300–350	9	200
XAD-4	Styrene divinyl benzene	850–950	5	200
Polyurethane foam	Polyurethane			

The resins Tenax GC and XAD-2 are general adsorbents in the sampling of organic compounds, and possess comparable properties. For its heat resistance, Tenax GC has been considered better in cases where the sample treatment involves thermal desorption. Where compounds are to be examined using solvent extraction, XAD-2 resin has been recommended. Polyurethane foam is suitable as an adsorbent in sampling of organic compounds having high boiling temperature (250–300°C).

5.8.3. PAH Compounds

Extraction from Particles

The efficiency of the extraction of PAH compounds from particle samples is affected by the physical and chemical character of the particles as well as the extraction solvent and the method of extraction. For example, different carbon matrices have a very different PAH affinity. In outdoor samples organic compounds are obviously for the most part adsorbed on the surface of the particles, from where it is easy to transfer them to the solvent. The extraction of fly ash, soot and diesel exhaust gas samples is considerably more difficult, and the yield of some compounds remains small, even though effective solvents were used.

The extraction solvent has to be selected according to the situation. Toluene has in many studies been regarded as the most efficient [97–99]. According to one study [98], toluene desorbs tetra- or polycyclic PAH compounds more effectively than methylene chloride. Aromatic solvents are, on the whole, considered more efficient than halogenated solvents. The minimal polarity of cyclohexane makes it selective to nonpolar compounds, such as normal PAH compounds.

Extraction from Adsorption Resin

Thermal desorption is particularly suitable for Tenax GC sorbent, but it has also been used for outdoor particle samples [10]. Using this method, PAH compounds can be sublimated directly from the sample to the column of a gas chromatograph. The advantage of heat desorption is the relatively short processing time, and that part of error and impurity sources (for example, glassware, solvents and elaborate fractioning) of the extraction and fractioning method can be avoided.

For the solvent extraction of the PAH compounds from the sorbent, the solvent should be the same as that used to clean the sorbent before the sampling. Some researchers recommend methylene chloride as the primary extraction solvent for XAD-2, because it has not been observed to affect the resin or its properties. According to other researchers, toluene is suitable for both Tenax and XAD-2.

Ethyl acetate and methanol are applicable only for the treatment of Tenax. PAH compounds have been extracted from polyurethane foam using e.g. methylene chloride, acetone and cyclohexane.

Separation and Fractioning

The organic material extracted from a flue gas or particle sample usually contains many different organic compounds [10], such as aliphatic hydrocarbons, polycyclic aromatic hydrocarbons and their oxygen, sulphur and nitrogen derivatives, organic oxygen compounds and other compounds. The complexity of the samples makes it more difficult to identify the PAH compounds and determine them by using conventional analysis methods. Therefore, they normally have to be separated from other compounds, and if one wants to analyse also the PAH derivatives, they have to be divided into several fractions.

The most common cleaning methods for samples containing PAH compounds are liquid–liquid extraction, column chromatography and thin-layer chromatography.

For the fractioning of PAH compounds, the separation into acid, neutral and alkaline parts using the liquid–liquid extraction, and further the fractioning of the neutral part by using column chromatography, has generally been used. In the past few years, the researchers have begun to use high resolution or high performance liquid chromatography (HPLC) for the fractioning of organic samples.

Analysis

Polycyclic organic compound groups differ so much from each other that usually the fractions obtained must be analysed separately [10]. It is not appropriate to use the same column and the same detection method for all PAH compounds. Furthermore, particularly in HPLC fractioning, disturbing aliphatic compounds and compounds belonging to other PAH derivative groups remain in the sample, which means that selective detection is needed for complicated samples. The most common analysis methods are gas chromatography, gas chromatography with mass spectrometry, and in some cases HPLC.

In gas chromatography, capillary columns have to be used, because there is such a great amount of compounds and their isomers to be separated.

Most of the sample fractions are so complex that polycyclic organic matter (POM) compounds cannot be identified only by gas chromatography. The components separated by a gas chromatograph can be identified based on mass spectra. Ordinary mass spectrometry is discussed further below in connection with analysis techniques. When using the selected ion monitoring technique (SIM), only ions of compounds having a certain molecular weight are selected for examination, in which case disturbance effects of other compounds can be eliminated. The mass spectra of the structure isomers of PAH derivates are, however, nearly similar, and therefore identification is possible only by simultaneously determining the chromatographic retention times.

HPLC can, in addition to fractioning, be used for the analysis of POM compounds. It is particularly suited for the determination of macromolecular compounds which are practically nonvolatile, since they can be separated at room temperature. The decomposition of compounds or the fact they may remain nonvaporised, which are problems in the gas chromatography, do not then affect the determination.

5.8.4. PCDD and PCDF Compounds

Extraction

For the determination of the concentration of chlorinated aromatic compounds, the sample taken from the flue gas is normally divided at least into three partial samples: fly ash sample collected on the filter, compounds concentrated from the gas phase in the adsorbent or absorbent, and water condensed in the cooler. The compounds to be studied from these partial samples are extracted using suitable organic solvents. The extracts obtained from the partial samples in ordinary emission measurements must be combined before further processing, because the distribution of compounds in the particle and gas phases under actual conditions is not known [100].

The extraction of chlorinated dioxins and furans from fly ash requires effective methods, for the concentrations of compounds to be studied are low in ash, and the compounds are tightly adsorbed on fly ash particles.

Acid treatment and extraction using toluene and benzene have been found applicable methods for the extraction of PCDD and PCDF compounds from fly ash. In acid treatment, the fly ash sample is elutriated in the acid, most commonly in hydrochloric acid which is left to act for an hour. After this the mixture is filtered, the ash sample is washed with water and dried. During acid treatment, the structure of the fly ash spherules is broken and the extraction solvent becomes more effective.

The extraction of chlorinated dioxins and furans from resins used as adsorbents is considerably easier than from fly ash. The resins are usually extracted in a Soxhlet extractor with toluene, but also benzene and hexane are used. In Germany, the PCDD and PCDF compounds in the gas phase are absorbed into an organic solvent, ethoxy ethanol, from where the compounds are extracted by methods of liquid–liquid extraction into toluene. The PCDD and PCDF compounds, which have been washed from the flue gas flow with the condensate, are extracted from the water phase most commonly by hexane.

Clean-Up and Fractioning

The raw extracts obtained in the first stage of the sample processing contain hundreds of organic compounds [10]. The analysis of chlorinated dioxins and furans occurring in small concentrations requires that the unwanted components are removed from the samples.

The methods employed in the clean-up of the samples are divided into two main groups. In the cleaning by column chromatography, the PCDD and PCDF compounds are separated into their own fractions, from which all compounds in question are analysed. In the cleaning by liquid chromatography, the dioxins and furans are separated from unwanted compounds, and are additionally divided into several fractions according to the degree of chlorination.

Analysis

Despite of effective pretreatment methods of samples, the analysis of the PCDD and PCDF compounds is based on sensitive equipment having good resolution. The most commonly used analysis apparatus is a capillary gas chromatograph combined with a mass-specific detector, or a mass spectrometer. When wishing to determine very low concentrations, and when the matrix is difficult, a capillary gas chromatograph is needed, from which the sample is conducted to a high-performance mass spectrometer, or a tandem mass spectrometer which is composed of two mass spectrometer units.

Of dioxins and furans, the supertoxin 2378 TCDD, which is a tetrachlorinated dioxin, is generally determined isomer-specifically. Its detection limit is < 1ppb.

5.8.5. PCB Compounds

The PCB compounds include a total of 209 chlorinated biphenyl compounds [10]. There can be up to ten chlorine atoms in a molecule.

For the extraction of PCB compounds from fly ash and adsorption resins, hexane, benzene, toluene and dichloromethane, for instance, are used. After the extraction the sample is cleaned of unwanted compounds. The sample may, for example, be treated with sulphuric acid, which oxidises other chlorine compounds than chlorinated aromatics in the sample, and cleaned by column chromatography.

The PCB concentrations of the samples are normally analysed by the gas chromatography using a packed column or a capillary column. The detector used is either an electron capture detector or a mass spectrometer.

5.8.6. Chlorinated Benzenes and Phenols

The chlorobenzenes, which total 12 different compounds, are determined by means of gas chromatograph plus electron capture detector apparatus, or gas chromatograph plus mass spectrometer equipment. The determination can take place, for example, from the same extract as the determination of the PCB compounds.

For the extraction of the chlorinated phenols, of which there are 19 in total, hexane and ether acidified by sulphuric acid are used, for example.

Chapter 6

Optical Spectroscopy in Emission Measurements

6.1. Infrared Spectroscopy

6.1.1. Energy Levels of Molecules: Vibration Spectra

The energy of molecules is the sum of translational, rotational, vibrational and electronic energies:

$$E_{mol} = E_{trans} + E_{rot} + E_{vib} + E_{el}$$

From the point of view of the molecular spectra, translational energy E_{trans} is not of considerable importance, and can be neglected [101]. Rotational energy can have some significance, above all in small molecules in a gaseous state, which can rotate relatively freely.

The electrons associated with the molecules and which form the electronic energy, are of three types: electrons that belong to one atom, electrons that are associated with two adjacent atoms, and electrons that are associated with more than two atoms. The influence of the first type of electrons, or the inner shell electrons, on the electronic transitions in question, which take place between the electronic states of the outer electrons and which could absorb light, is negligibly small and can be omitted.

The rest of the electrons can participate in the electronic transitions between rotational and vibrational energy states, which can be induced by ultraviolet or visible light. Infrared radiation is associated with a lower energy, which cannot normally induce these types of transitions.

In the process of the absorption of infrared radiation, the most significant interaction, associated with the exchange of energy, can be considered to happen between the radiation and the atomic vibrations in the molecules.

6.1.2. Degrees of Freedom

Each atom of a monatomic gas has three degrees of freedom, one for each spatial direction. A diatomic molecule has, in addition to these three translational degrees of freedom, two degrees of freedom associated with its rotation. Generally, if the molecule is not linear, there are three rotational degrees of freedom. If the atoms of a diatomic molecule can vibrate in an interconnected way, this forms an additional degree of freedom. A diatomic molecule thus has six (3 × 2) degrees of freedom. Degrees of freedom are illustrated in Fig. 6.1.

In a three-dimensional space an N-atomic molecule has $3N$ degrees of freedom or $3N$ possible motions. Of these, three are translations and three rotations around different axes. However, a linear molecule has only two rotations. The other $3N - 6$ motions are vibrations ($3N - 5$ with a linear molecule).

6.1.3. Interaction between Matter and Radiation

Energy is conserved in the interaction between matter and radiation. In the absorption of radiation, a quantum of energy of magnitude hv (h is Planck's

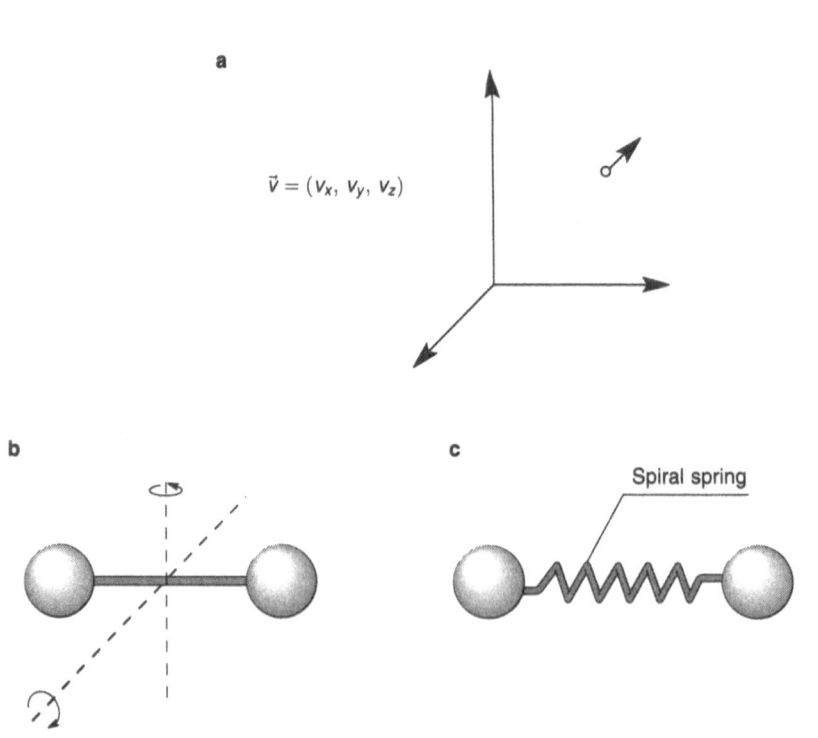

Figure 6.1. a Translational degrees of freedom of a monatomic gas. **b** Rotational degrees of freedom of a diatomic gas [102]. **c** Vibrational degree of freedom of a diatomic molecule.

constant and v is the frequency) is transferred to the molecule. The precondition for the absorption is that there are energy states in the molecule, and that the energy difference between each is exactly hv.

The state of a molecule is defined by a group of quantum numbers, which describe the rotational, vibrational and electronic state of the molecule. As an absorption or emission takes place, a change in some of the quantum numbers occurs simultaneously. Still only certain transitions are possible. These transitions must follow the so called selection rules.

Based on the above, it can be supposed that the interaction between the matter and the radiation depends on the amount of energy associated with the radiation or the wavelength of the radiation. The interaction between the matter and the radiation thus includes phenomena from the nuclear and inner shell electron transitions to molecular rotations and translations. Different phenomena associated with the interaction between radiations of different energies and the matter are presented in Table 6.1.

Table 6.1. Ranges of the electromagnetic spectrum and the corresponding energy transitions [74]

Range	Approximate limit of the		Phenomenon
	wavelength λ	frequency v (Hz)	
γ-rays			Nuclear transitions
	10 pm	3.0×10^{21}	
X-rays			Transitions of the inner shell electrons
	10 nm	3.0×10^{18}	
Vacuum ultraviolet			Removal of the outer shell electrons
	200 nm	1.5×10^{17}	
Ultraviolet			Transitions of the valence electrons
	380 nm	7.9×10^{14}	
Visible light			Transitions of the valence electrons
	780 nm	3.8×10^{14}	
Near infrared			Molecular vibrations
	2.5 μm	1.2×10^{14}	
Mid-infrared			Molecular vibrations
	50 μm	6.0×10^{12}	
Far infrared			Molecular vibrations
	1 mm	3.0×10^{11}	
Microwaves			Molecular rotations and electron spin resonance
	100 mm	3.0×10^{9}	
Radiowaves			Nuclear magnetic resonance

6.1.4. Infrared Absorption

In infrared spectroscopy, we are dealing with molecular vibrations and rotations. The more significant and, as regards the characteristic features of the molecular spectra, the more essential part, is the vibration between the atoms in the molecule. When transferring from one vibrational state to another, the molecule can absorb infrared light. From the combination of absorptions thus occurring, an infrared spectrum, characteristic of the molecule, is generated. This spectrum can be used to identify matter, to analyse it quantitatively, and to characterise its structure. In the transition of a molecule from one vibrational state to another, a change in both the vibrational and translational energy states normally takes place.

6.1.5. Vibration Frequency

In a molecule comprising two (or more) atoms, the atoms are in permanent vibrational motion in relation to one another. Vibration is one of the forms of energy, and it depends, e.g. on temperature. Each bond has its characteristic vibration frequency, and it can therefore absorb radiation of just a certain wavelength. From this, an absorption spectrum characteristic of each molecule, also called a "fingerprint spectrum", originates.

The simplest molecule, the atoms of which can vibrate, is a diatomic molecule. When building up the chemical bond of such a molecule, two forces, the attractive and repulsive forces, will be set in balance. The essential features of a molecular spectrum are revealed by the absorption spectrum of a diatomic molecule.

A diatomic molecule can vibrate only in the direction of bond between the atoms, like two objects linked together by a spiral spring. When the amplitude of the vibration is within certain limits, it is harmonic, and Hook's law applies. Then

$$\tilde{v} = \frac{1}{2\pi c} \sqrt{\frac{k}{\mu}} \tag{6.1}$$

where $\tilde{v}$ is the "wavenumber" of the vibration, c is the velocity of light, k is the "spring constant", which describes the force binding the atoms together, and μ is the reduced mass

$$\mu = \frac{m_1 m_2}{m_1 + m_2} \tag{6.2}$$

where m_1 and m_2 are the masses of the atoms (1 and 2).

By applying the value of k determined for a single bond ($5\,\mathrm{N\,cm}^{-1}$) and the value of the atomic mass unit ($1.66 \times 10^{-24}\,\mathrm{g}$) we obtain for the bond between carbon and hydrogen C—H, the reduced mass of which is approximately one:

$$\tilde{v} = 1302 \sqrt{\frac{3}{1}} \approx 2910 \, cm^{-1}$$

The absorption of the bond C—H in methane, for instance, is observed at $2915 \, cm^{-1}$, which is almost the same as the result of the approximate calculation performed above.

A diatomic molecule has thus only one vibrational motion, the stretching of the bond. The vibration of multiatomic molecules can, on the other hand, be very complex. In addition to the stretching, it contains bending vibrations (Fig. 6.2).

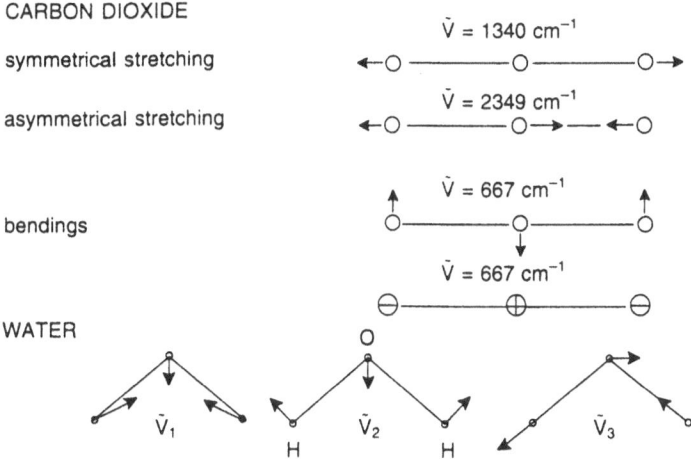

Figure 6.2. Scheme of the normal vibrations of carbon dioxide and water molecules [74]. Two bending vibration types are similar, but take place in planes rectangular to each other. One of them is indicated by + and −. It can be seen that the wavenumbers of the stretching vibrations are higher than those of bending vibrations. This reflects the fact that more force is normally needed to stretch the interatomic bond than to bend the bond angle.

6.1.6. Wavenumber

The wavenumber $\tilde{v}$ is defined as the reciprocal of the wavelength λ of the electromagnetic radiation, or $\tilde{v} = 1/\lambda$. It is generally used in infrared spectroscopy. A wavenumber can be expressed in units of, e.g. cm^{-1}.

6.1.7. Characteristic Frequencies

The infrared absorption spectrum is characteristic of each absorbing molecule, as can be deduced from the above. It has been noted, however, that many of the vibration frequencies of the molecules are due to small atomic groups. These frequencies are characteristic of atomic groups in question, and remain roughly the same, independent of the molecule in which they occur. These frequencies are called group frequencies.

In addition to the group frequencies, there are vibrations in all the molecules characteristic of these molecules, which depend on the space-geometrical factors as well as on the atomic masses and characteristics. Together with the group frequencies, the absorptions caused by these vibrations form the characteristic spectrum, based on which almost every molecule that absorbs infrared radiation can be identified.

In normal practice, the spectra contain additional transitions associated with vibration frequencies, which contain both the vibration and the rotation. The vibration frequencies are, however, hundreds to thousands times higher than the rotation frequencies, so that the frequency of the pure vibration differs only slightly from the vibration–rotation frequency. In the spectrum, the rotation causes frequencies on both sides of the the frequency of the pure vibration, forming a vibration–rotation band.

6.1.8. Rotation Frequencies

As an example of the quantitisation of the rotation energy and the absorption of the infrared light caused by it, we can consider the rotation of a molecule formed by two atoms (Fig. 6.3). Diatomic molecules, similar to linear molecules, have two equal principal moments of inertia. Ball-like spinning tops have three mutually equal principal moments of inertia.

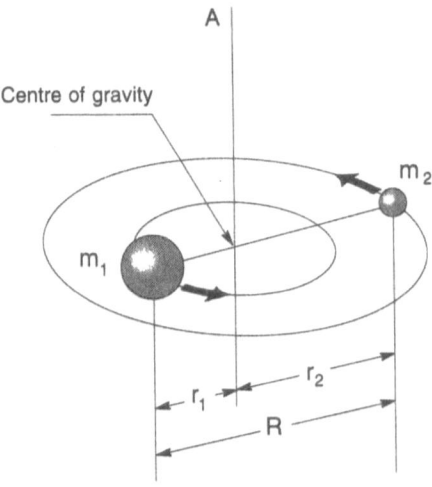

Figure 6.3. Rotational motion of a diatomic molecule [102].

The moment of inertia I of a diatomic molecule is [102]

$$I = \frac{m_1 m_2}{m_1 + m_2}(r_1 + r_2)^2 \qquad (6.3)$$

The rotation energies of linear and ball-like spinning top molecules can be obtained from the equation [102]

$$E_r = \frac{h^2}{8\pi^2 I}J(J+1) \qquad (6.4)$$

where h is Planck's constant and J is the rotational quantum number $J = 0, 1, 2, \ldots$ and it can only change by steps of $\Delta J = \pm 1$. The energy change of the transition $J \rightarrow J - 1$ is

$$\Delta E = \frac{h^2}{4\pi^2 I}J \qquad (6.5)$$

The rotational energy levels of a rigid diatomic molecule, together with the corresponding spectrum for transitions $\Delta J = 1$ are presented in Fig. 6.4.

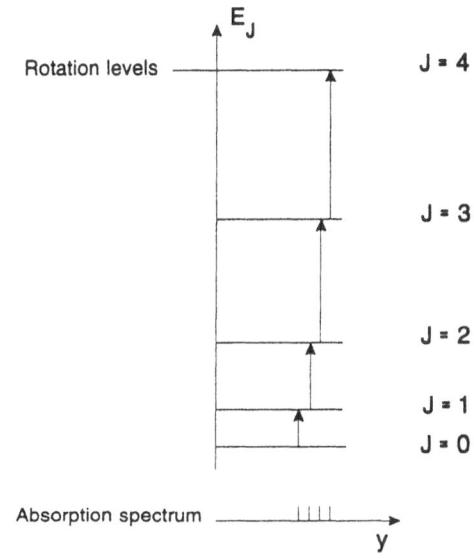

Figure 6.4. Rotational energy levels of a rigid diatomic molecule and the corresponding spectrum for transitions $\Delta J = 1$ [102].

6.1.9. Rotational Spectra of Polyatomic Molecules

The moments of inertia of molecules naturally vary according to their shapes. In Table 6.2 some molecules are categorised according to their shape.

Table 6.2. Categorisation of some molecules according to their moments of inertia [74]

Moments of inertia	Type of molecule	Examples
$I_b = I_c$, $I_a = 0$	Linear	HCN
$I_a = I_b = I_c$	Ball-like spinning top	CH_4, SF_6, UF_6
$I_a < I_b = I_c$	Oblong symmetrical spinning top	CH_3Cl
$I_a = I_b < I_c$	Flattened symmetrical spinning top	C_6H_6
$I_a \neq I_b \neq I_c$	Asymmetrical spinning top	CH_2Cl_2, H_2O

6.1.10. Vibration–Rotation Spectra of Diatomic Molecules

When investigating the molecular spectra of gaseous species using high resolution spectrometers, their spectral lines can be noticed to have been formed by a large number of components near each other. These spectra are therefore called band spectra, whereas spectra obtained from atoms are called line spectra [74]. The fine structure of the spectra of molecules is due to the fact that, associated with the vibration transitions, excitations of rotational motions are generally also created. In Fig. 6.5, a high resolution spectrum obtained from hydrogen chloride gas is presented.

The excitation of the rotational transitions in connection with vibrational transitions can often be considered, based on the fact that the energies asssociated with the vibrational transitions are so high compared with the rotational transitions, that there is excess energy to be transferred to the rotational transitions. Associated with the change in the vibration, there is also a change in the interatomic binding, which influences the moment of inertia of the molecule, and is thus linked together with a change in the rotational state. It has to be noted, however, that when J increases, the differences between the rotational energy levels also increase. The general precondition for absorption is that the energy quantum absorbed from the radiation coincides with the energy difference between the energy states, whether this difference be based on either vibrational or rotational states.

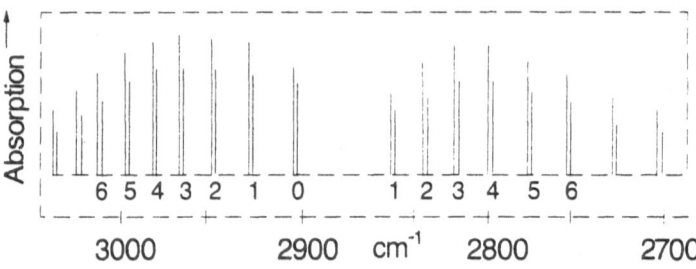

Figure 6.5. High resolution spectrum of HCl gas. Dual lines are due to the different isotopes of chlorine (75% ^{35}Cl and 25% ^{37}Cl) [74].

For the rotational change associated with a vibrational transition, the selection rule $\Delta J = \pm 1$ applies. Sometimes $\Delta J = 0$ is also possible.

Based on the harmonic oscillator and the rigid rotator models, the energy levels of a vibrating and rotating molecule can be presented as follows [74]:

$$E_{vJ} = \left(v + \tfrac{1}{2}\right) h v_e + hc B_v J(J + 1) \tag{6.6}$$

where v is the vibration quantum number and J is the rotation quantum number. The subscript e refers to the equilibrium state. B_v is a constant associated with the rotation and dependent on the vibration quantum number v. Vibrational and rotational transitions are thus linked together. When a vibrational transition happens, not only does v change to $(v + 1)$, but also J changes by either $+1$ or -1 (or sometimes $\Delta J = 0$). The vibrational absorption can thus be divided into three groups of lines. These are called branches and are indicated by P, Q and R. For instance, the lines of the P-branch correspond to transitions $\Delta v = 1$ and $\Delta J = -1$. If the molecule is originally at the rotational state J, the transition is $J \rightarrow J - 1$, and the total change in the energy is

$$\Delta E_{vJ} = h v_e - 2 B_v hc J \tag{6.7}$$

The origin of the vibration–rotation spectrum of a diatomic molecule is presented in Fig. 6.6.

Vibration spectra can be interpreted and used for chemical analysis by taking into account some simplifying regularities associated with the molecular structures. These include

1. The molecular symmetry.
2. Group frequencies mentioned before. As an example, the following wavenumbers corresponding to group frequencies can be mentioned:

O–H	3650 cm^{-1}
N–H	3450
–C–H	3300
$= CH_2$	3100
$-CH_3$	2950
$>C=O$	1850
	1650
CH_2	1450

3. Effect of the isotope on the wavenumber:

$$4\pi^2 c^2 \tilde{v}^2 = f/m$$

This shows that the wavenumber $\tilde{v}$ is proportional to the reciprocal of the square root of the effective mass.

In addition to the values given earlier, some group frequencies and types of the vibrations, are presented in Table 6.3.

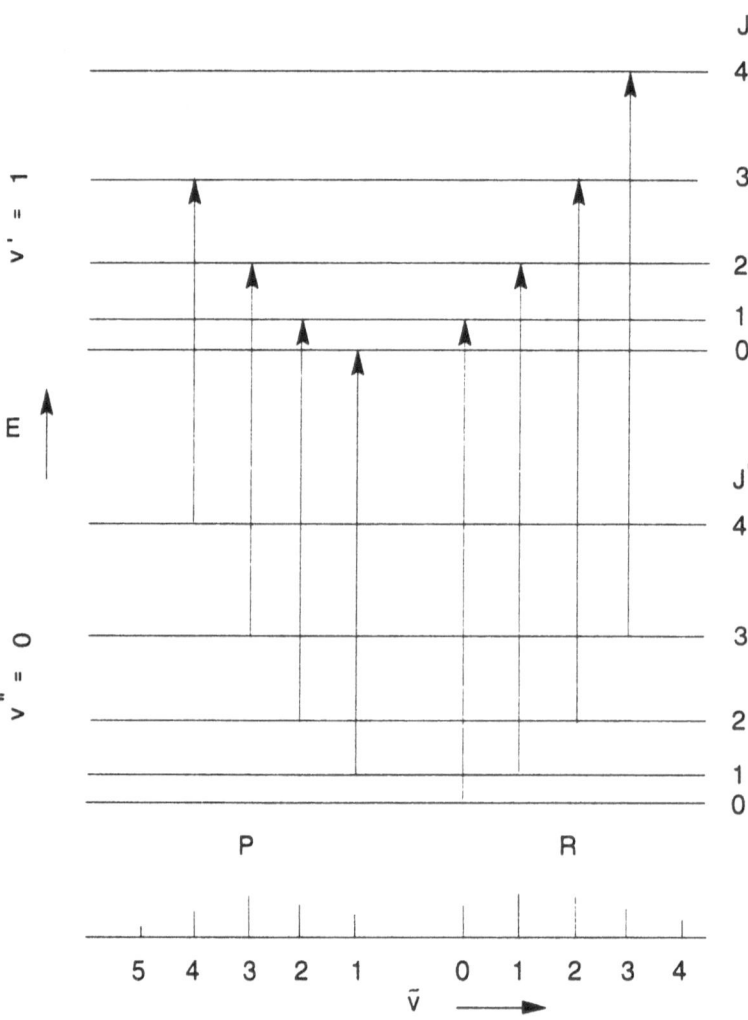

Figure 6.6. Scheme of the formation of the vibration–rotation spectrum of a diatomic molecule [74].

Vibrational spectra contain principally the basic bands of the vibration, that are observed at the frequencies (f) of the vibrations taking place in the molecules (or at the wavenumbers $\tilde{v}$) [104]. A diatomic molecule, for instance, has only one vibration. A molecule containing N atoms has

- $3N - 6$ vibrations, if the molecule is not linear and
- $3N - 5$ vibrations, if the molecule is linear.

In the vibration spectra, there can also appear harmonic vibrations or combination bands, generated by multiples and sums of the basic vibrations

Table 6.3. Group frequencies of some compounds in infrared spectra [103]

Compound group	Typical bands (cm^{-1})	Absorption basis[a]
Carbonyls	1809–1686	C=O (s)
Aromatic compounds	3094–3013	C–H (s)
	1643–1586	C=C (s)
	1562–1454	C=C (s)
Carboxylic acids	1836–1759	C=O (s)
	1123–1096	C–O–C (s)
Aldehydes	2739–2670	C–H (s)
	1732–1701	C=O (s)
Nitrogen compounds	2824–2782	N–H (s)
	1628–1601	N–H (b)
	1589–1535	N–O (s)
Phosphor compounds	1072–1018	P–O–R (s)
Ethers	1146–1099	R–O–R (s)
Alcohols	3720–3550	O–H (s)
	1084–1003	C–O (s)
Esters	1767–1736	C=O (s)
	1292–1262	C–O–R (s)
	1238–1215	C–O–R (s)
	1177–1146	C–O–R (s)

[a] s: stretching; b: bending.

which are normally much weaker than the basic vibrations. Anharmonicity causes the frequency of these vibrations to differ from the precise multiple or sum of the basic vibrations.

6.1.11. Precondition for Infrared Absorption

The precondition for the absorption of the infrared light is that there is a permanent dipole moment in the molecule, or that the dipole moment is built up as the atoms vibrate. This condition is met in heteroatomic molecules, i.e. in molecules which consist of different atoms. On the other hand, the molecules of, e.g. gases, which are formed by two similar atoms, do not absorb infrared light. That is why oxygen and nitrogen, for instance, cannot be detected or analysed by means of infrared absorption.

6.1.12. Infrared and Raman Spectroscopies

Vibration spectra can be interpreted using infrared absorption spectroscopy or Raman spectroscopy. These two techniques complement each other well. As became apparent earlier, the vibrational frequencies, and thus the wavelength (and the wavenumber) of the light absorbed, depend on the strength of the bonds between the atoms. These strengths of the bonds, in turn, are characteristic of the compounds built up by different atoms, and thus the vibration spectra can yield information on the concentrations of the compounds.

Information from vibration spectra can be obtained by using either infrared absorption spectroscopy or Raman scattering spectroscopy. In these spectra, only a limited collection of basic vibrations normally occurs, based on selection rules. The dominating selection rules are:

- Infrared absorption. The vibration of the atoms in the molecule must be associated with a change in the electrical dipole moment.
- Raman scattering. The vibration of the atoms must be associated with a change in the polarisability.

6.2. Gas Measurements by Infrared Absorption Techniques

Infrared spectroscopy can be considered to be one of the best analytical methods to describe the molecular structures of compounds. It can also be used for the measurement of the concentrations of gases as well as for the investigation of molecular structures.

The state of the matter to be analysed influences the infrared spectrum. The influence mainly concerns the rotational peaks, so that the fine structure caused by them becomes apparent in the spectra of the gaseous state. This is because the molecules can vibrate and rotate most freely in the gaseous state.

As was stated earlier, based on the atomic vibrations, molecules can absorb infrared radiation only if the dipole moment set up by the vibrating atoms changes during the vibration. This does not happen in molecules comprising two identical atoms. As far as gas measurements are concerned, this means that the most important gaseous components of the air, oxygen and nitrogen, do not absorb infrared radiation. This makes it possible to measure, based on the infrared absorption, concentrations of other gases in air or in emission gases.

Almost all gases can be measured using infrared techniques. Absorption spectra are most often utilised. Molecules comprising two different atoms produce one basic band, whereas polyatomic molecules produce two or more significant spectral bands.

6.3. Measurement of Emission Gases Based on Infrared Absorption

The concentrations of the emission gases to be measured are higher than, e.g. the concentrations of the impurity gases in the atmosphere. Emission gases can thus be measured using instruments not necessarily meeting the highest resolution and sensitivity requirements. This enables use of a wide-band infrared source and selection of the measurement band using filters. Such instruments are called "nondispersive" infrared (NDIR) instruments. Non-dispersive methods use filters or cells, filled with the gas to be measured, as selective filters. Bandpass filters pass the wavelength bands studied and they can be placed on either the source or the receiver side [104]. In the following, two examples of NDIR analysers are presented.

A measuring arrangement based on the use of filters is shown in Fig. 6.7. Two narrow wavelength bands, so-called absorption and reference bands, are selected by filters. These bands are selected so as to obtain high and low absorption, at absorption and reference bands, respectively. The filters are mounted on a rotating disc, and the intensities are measured synchronously. Then the ratio of the intensities is dependent on the concentration of the absorbing gas. The lowest concentration that can be measured, is determined by the strength of the band and by the ratio of the bands used and, of course, by the length of the measuring path. The interference caused by other gas components can be reduced by measuring the concentration of the interfering gas at another band, and correcting the measured intensities mathematically. The measurement based on the ratio of the intensities is insensitive to variations in the intensity of the light source. On the other hand, error in the result of the measurement can be caused by variations, which include dependence on the wavelength. Such factors include:

- Variations in the temperature of the source
- Variations in the wavelength-response of the detector
- Attenuation caused by dust on the measurement path and the fouling of the optical components.

To reduce the influence of the error factor, it is advantageous to choose the measurement and reference bands near each other.

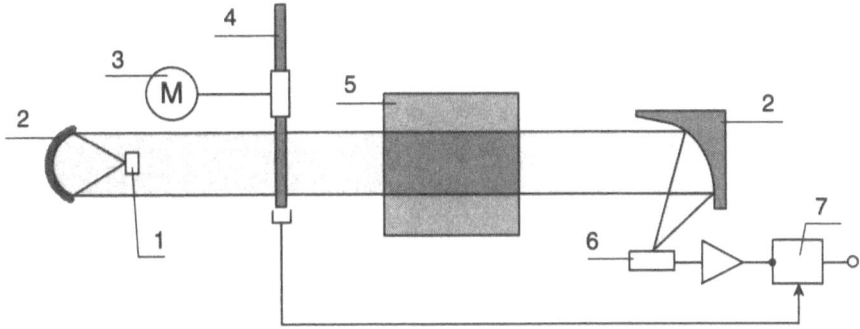

Figure 6.7. An NDIR analyser based on the interference filters [104]. 1, infrared source; 2, mirror; 3, motor; 4, filter disc; 5, absorbing gas; 6, detector; 7, synchronous indication.

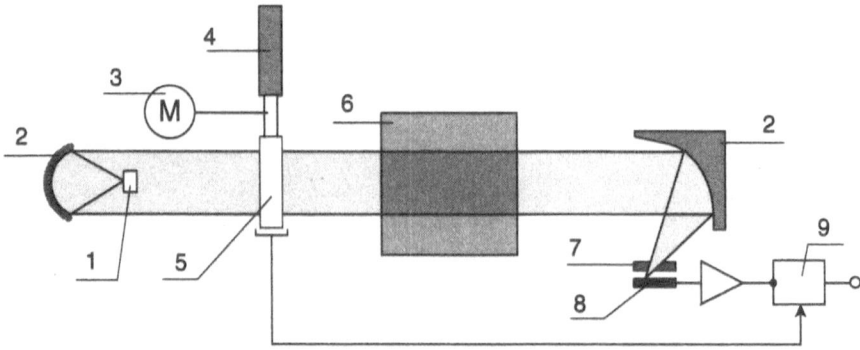

Figure 6.8. An NDIR analyser based on the correlation method [104]. 1, infrared source; 2, mirror; 3, motor; 4, grey filter; 5, correlation cell; 6, absorbing gas; 7, wavelength filter; 8, detector; 9, synchronous indication.

In the correlation method, the light path is crossed, in turn, by a cell containing a high concentration of the gas to be measured (correlation cell), and a cell containing gas that does not absorb at the measuring band (e.g. nitrogen, grey filter, Fig. 6.8). The correlation cell causes peak absorption at the absorption band of the gas to be measured. The correlation cell thus filters this band "black", so that the absorption by the gas to be measured, in the measurement cell, no longer affects the reference signal. The concentration of the gas to be measured does affect, however, the intensity of the light passing through the nitrogen cell and the measurement cell. This type of instrument can be used for the measurement of several gas components, if the rotating wheel is provided with several cells containing different gases.

In the correlation method, the measurement and the reference signals are obtained from the same wavelength band. The variation in the temperature of the source, or other factors causing changes in the wavelength distribution, thus do not disturb the measurement.

6.4. FTIR techniques in Measurement of Gases

Fourier transform infrared (FTIR) techniques offer a high resolution, and are therefore very useful in the measurement of gases. Compared with most other methods, their main advantage is the capability of analysing several gases. As noticed earlier, gas analysers can normally be used to measure only one gas component. Still another advantage compared with the spectroscopic infrared method is that compounds disturbing the measurement can be recognised from the spectrum, and their concentration can be determined, if needed [105].

In the infrared spectra of most molecules, there are specific features which enable these molecules to be identified. The differences are most pronounced in the case of small molecules, and there are precise details in the absorption bands. Using high resolution instruments, these compounds can be separated from each other and their concentrations determined, even though the spectra of different compounds partly overlap. Using these instruments, even the absorption caused by water vapour can be separated, despite its interference at several points of the spectrum.

As was stated earlier, infrared techniques can be applied to the measurement of the concentration of very many gases. In flue gas analysis, these techniques form a practically continuous analysis and measurement method for several gases. They have been applied, for instance, in the measurement of the concentration of the following gas components from flue gases: CH_4, C_2H_2, C_2H_4, C_2H_6, CO_2, CO, SO_2, NO, NO_2 and N_2O [105, 106]. In the measurement of N_2O, FTIR analysis is virtually the only reliable continuous method.

For an example of infrared absorption spectra produced by gases, see Sect. 6.4.6 (Fig. 6.16 [105]). The measurement was performed in a combustion experiment, in which peat was used as fuel. It appears from the figure, that the spectrum of carbon monoxide, which is a diatomic gas, contains only one basic band, whereas the spectra of polyatomic molecules contain several spectral bands.

6.4.1. FTIR Spectrometer

The operation of a dispersive infrared spectrometer utilising the Fourier transform is based on the interference of two light beams, passing through different paths, and the interpretation of the interferogram obtained by a detector from the interferences of these two light beams. The two beams originate from the same source, but the light is divided into two parts using a semi-transparent ("half-silvered") mirror or a beamsplitter. The most commonly used technique is based on the Michelson interferometer. A scheme of an FTIR spectrometer is presented in Fig. 6.9. As can be noticed, one part of the light is reflected from a fixed mirror, whereas the other part is reflected from a moving mirror, by means of which the interference is formed. The method is thus based on the coherence of two light components coinciding at the beamsplitter, the resulting interference spectrum being registered.

To be able to interpret the interferogram, the exact position of the mirror must be known. The position is measured using monochromatic light obtained from the reference laser. As will be noticed later, the accuracy of the movement of the mirror is an important factor regarding the resolution of the instrument. It is the basic factor which determines how detailed a spectrum can be formed with the instrument. The scheme of an FTIR analyser is presented in Figure 6.10. An arrangement used to eliminate the error caused by the fluctuation of the mirror is also shown in the figure. This arrangement is called a cube corner.

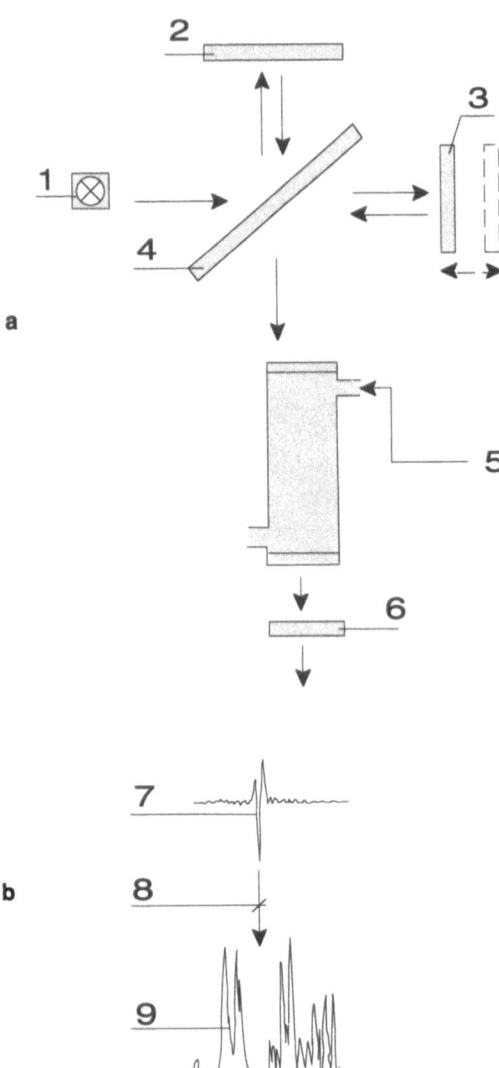

Figure 6.9. a Operation principle of an FTIR spectrometer (Michelson interferometer). 1, infrared light source; 2, fixed mirror; 3, moving mirror; 4, beam splitter; 5, sample gas; 6, detector. **b** Mathematical treatment of measurement result: 7, interferogram; 8, Fourier transform; 9, spectrum.

The application of the FTIR techniques and the associated separation of spectra from one another as well as the subtraction of the background spectra require laborious calculations to be performed. Computers are therefore used as integral parts of FTIR instruments. To measure the position of the moving mirror, helium–neon lasers are used in FTIR instruments. It is necessary to measure accurately the position of the mirror so as to obtain the real interferogram and the real spectrum derived therefrom to describe the energy

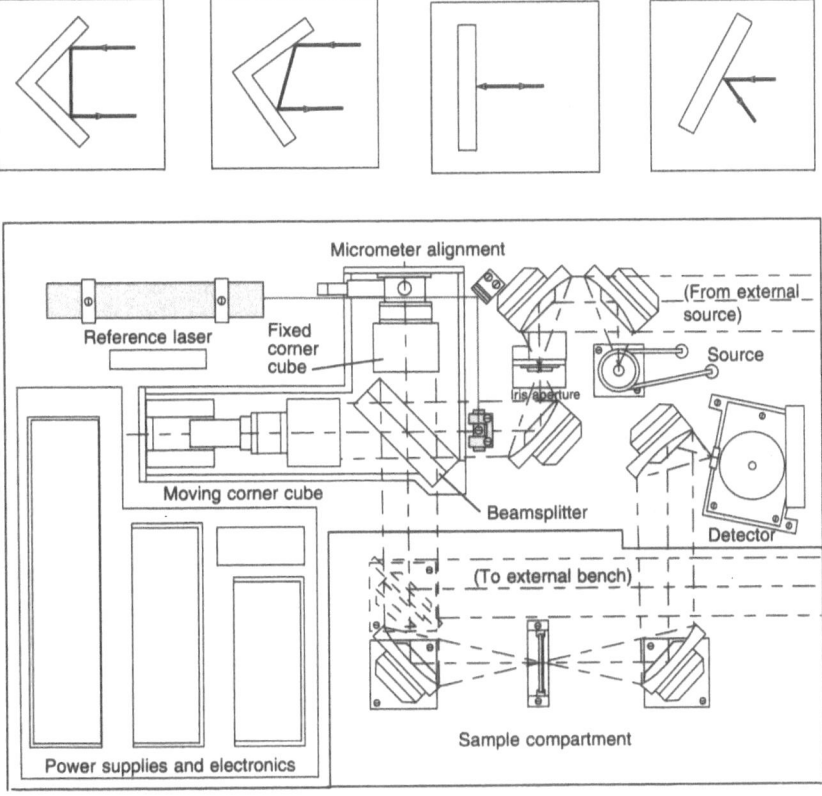

Figure 6.10. Construction of an FTIR spectrometer and the arrangement of the mirrors [107].

of light as a function of the wavenumber. The accuracy must be within a fraction of the wavelength.

A basic practical FTIR instrument consists of the following units:

1. Optical system, which produces the interferogram.
2. Computer, which controls the optical system, stores the interferogram, calculates the fast Fourier transform, and performs the other necessary calculations.
3. Program. The performance of the instrument is based on the usefulness and sophistication of the program.
4. Display unit.

Some interferograms, together with transmission spectra derived from them using the Fourier transform, are presented in Fig. 6.11.

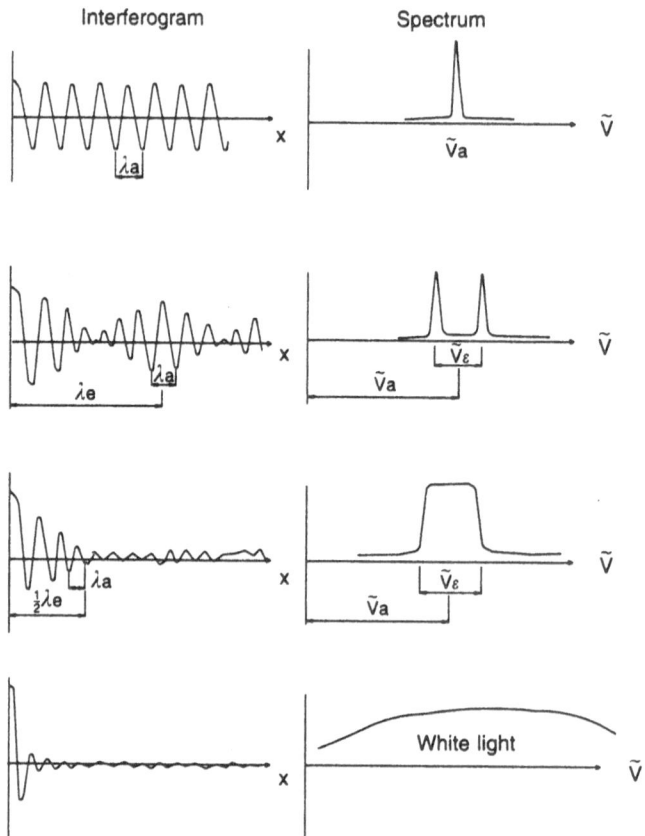

Figure 6.11. Different interferograms and corresponding energy spectra [108].

6.4.2. Properties of Fourier Transform Spectrometers

The signal reaching the detector of an FTIR spectrometer is thus not a normal spectrum, but the Fourier transform of a spectrum. It must be converted into the form of a spectrum again using a Fourier transform. Only then can a gas analysis (or some other analysis) be performed. The signal reaching the detector, dependent on the position of the mirror, contains a great number of wavelengths at each moment. Compared with other dispersive methods, FTIR techniques offer an excellent signal-to-noise ratio. This is because [72,108]

- The light energy utilised by the FTIR analyser is high, as no narrow slits are needed to guide the light.
- All the wavelengths of the light reach the detector throughout the duration of the measurement.

The slits used in other dispersive analysers reduce considerably the light energy available for the analysis, and each wavelength to be analysed is only observed for a very short moment, as it strikes the detector through the slit.

When the light power involved is small, the advantages mentioned above make the FTIR analyser especially efficient compared with other dispersive infrared analysers. Such cases include the measurement of high-resolution spectra, rapid recording of spectra, and the measurement of the infrared absorption. The last point is important when considering the use of FTIR analysers for gas analyses.

Recent work in the research and development of infrared spectrometers is aimed at constructions which could use linear arrays of sensors to detect different wavelengths of the light. The movement of the mirror could then be avoided.

6.4.3. Construction of Spectrum

In the following, the optical path difference between the beams coming from the fixed and the moving mirror is designated by x, and $\tilde{v} = 1/\lambda$ is the wavenumber of the radiation. We then obtain for the intensity of light at the detector

$$I(x, \tilde{v}) = \frac{1}{2} B(\tilde{v})(1 + \cos(2\pi\tilde{v}x)) \tag{6.8}$$

in which $B(\tilde{v})$ is the intensity of light coming to the beamsplitter. The total intensity arriving at the detector is

$$I(x) = \int_0^\infty I(x, \tilde{v}) \, d\tilde{v} \tag{6.9}$$

As we are interested in the alternating component, we take only this component into account obtaining

$$\bar{I}(x) = \int_0^\infty \frac{1}{2} B(\tilde{v}) \cos(2\pi\tilde{v}x) \, d\tilde{v} \tag{6.10}$$

and, according to the theory of the Fourier transform

$$\bar{B}(\tilde{v}) = 8 \int_0^\infty \bar{I}(x) \cos(2\pi\tilde{v}x) \, dx \tag{6.11}$$

The interferogram, which is in distance-space (x-space), has been converted into a spectrum in the wavenumber-space ($\tilde{v}$-space). If the optical path difference is changing at a constant rate of v, the detector will experience an alternating signal of frequency

$$f(\tilde{v}, v) = v\tilde{v} = \frac{v}{\lambda} \tag{6.12}$$

In the commonly used Michelson interferometer, v is twice the speed of the moving mirror. This speed can typically be about $0.5\,\text{cm s}^{-1}$, and $\tilde{v}$ can be from $400\,\text{cm}^{-1}$ to $4000\,\text{cm}^{-1}$. The frequency f is then from $0.2\,\text{kHz}$ to $2\,\text{kHz}$.

Significant factors influencing the choice of the speed of the change of the optical path difference and thus the speed of the mirror are:

1. Speed of the infrared detector. Photoelectric detectors, such as HgCdTe are fast (about $10\,\text{kHz}$), and TGS is slow (about $100\,\text{Hz}$).
2. Speed of the analogue-to-digital (AD) converter ($> 100\,\text{kHz}$).
3. Data collection capacity of the computer.

Factors 1 and 2 normally limit the highest frequency to about $10\,\text{kHz}$.

6.4.4. Resolution of Instrument and Maximum Optical Path Difference

The maximum optical path difference is $\Delta x_{max} = 2 \times$ (maximum mirror movement distance). The highest resolution is $\sigma\tilde{v} = 1/\Delta x_{max}$.

If, for instance, the maximum mirror movement distance is $1\,\text{cm}$, then $\Delta x_{max} = 2\,\text{cm}$ and the resolution is $\sigma\tilde{v} = 0.5\,\text{cm}^{-1}$. If the maximum mirror movement distance is $10\,\text{cm}$, the highest resolution is $0.05\,\text{cm}^{-1}$. High resolution thus requires large mirror movements with precision construction, which tends to make these instruments expensive. The highest resolution of the instruments used for emission measurements is normally about $0.5\,\text{cm}^{-1}$ (or $1\,\text{cm}^{-1}$).

6.4.5. Components and Materials of FTIR Instruments

Some materials used in the detectors of the FTIR spectrometers together with the useful detection ranges are presented in Table 6.4.

Table 6.4. Properties of infrared detectors [109]

Detector	Operation temperature	Detection range (cm^{-1})
TGS DTGS	Room temperature	5000–~30
PbS PbSe	Liquid nitrogen temperature	> 3000
InSb	Liquid nitrogen temperature	20000–1200
HgCdTe (MCT)	Liquid nitrogen temperature	3000–600
Bolometer	Liquid helium temperature (1.9 K)	450–10

[a] TGS: triglycine sulphate; DTGS: deuterated TGS; MCT: mercury cadmium telluride.

The DTGS detector is applicable to the mid-infrared range (4800–250 cm^{-1}), similar to the wide-band MCT. The narrow-band MCT can be used for high sensitivity measurements in the wavenumber range of 5000–750 cm^{-1} when using KBr or Ge beamsplitters, and in the range 7000–4000 cm^{-1} when using a quartz beamsplitter.

The indium–antimony (InSb) detector is applicable to the measurements in the near-infrared range (10000–1800 cm^{-1}). Additionally, silicon detectors (about 25000–8000 cm^{-1}) and germanium detectors doped with zinc (near-infrared) can be used.

Materials used for beamsplitters are listed in Table 6.5. Other materials used include Ge and KCl. The operation range of the Ge-beamsplitter is roughly the same as that of the KBr-beamsplitter, approximately 4800–400 cm^{-1}.

Table 6.5. Materials for beamsplitters [109]

Infrared range	Wavenumber range	Material
Near infrared	20000–3000	Quartz
	10000–2000	CaF_2
Mid-infrared	4000–400	KBr
	800–200	CsI
Far infrared	650–100	Mylar (and others)

For the windows of the infrared measuring instruments, the proper choice of material is also essential. Ordinary quartz and glass, for instance, cannot be used, as they absorb infrared frequencies. In Table 6.6, materials transparent to infrared light are listed, together with some characteristic properties.

Table 6.6. Window materials transparent for infrared light [110].

Material	Range (cm^{-1})	Solubility in water ($g\,dm^{-3}$)	Remarks
NaCl	650–50000	360	Cheap, cleaving and polishing easy
KBr	400–10000	540	Cleaving and polishing easy
CsI	200–10000	440	Soft, very expensive
CsBr	280–20000	1230	Soft, very expensive
CaF_2	1250–48000	0.017	More expensive than NaCl, polishing difficult, cannot be cleaved
BaF_2	1000–5000	1.7	Cannot be cleaved, more expensive than NaCl
AgCl	450–4000	–	Soft, light-sensitive (dark)
KSR-5 (TlBr-TlI)	250–20000	–	Expensive, poisonous
Polyethene	1–600	–	Especially for far infrared range
ZnS (Irtran)	715–17000	–	Polishing difficult, expensive

6.4.6. Gas Analysis by FTIR Techniques

Background Spectrum

The background spectrum is stored in the memory of the computer prior to the measurement of the sample. This is done by measuring the spectrum of the sample chamber without the sample gas, and its purpose is to exclude the influences of the measuring instrument and other gases apart from the one to be measured. Although the inner part of the spectrometer is continuously flushed by dry nitrogen or air, the signals reaching the detector from water and carbon dioxide can vary during a long measurement. Therefore the final result is formed by subtracting a background spectrum from the measured spectrum so that these two spectra are timely as near each other as possible [111]. Thus the influence of the background can be excluded. A background spectrum is presented in Fig. 6.12. It shows the general form of the spectrum due to the light source, and the absorption caused by water and carbon dioxide.

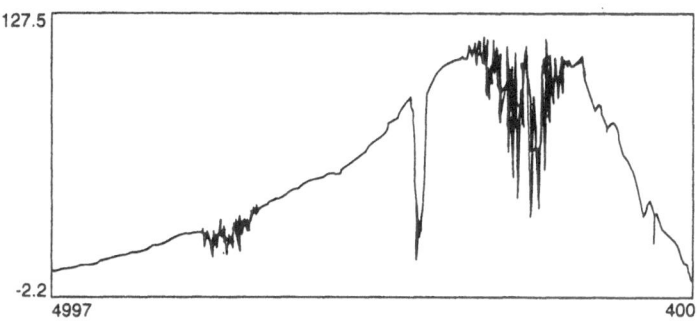

Figure 6.12.   Background spectrum [112].

Spectrum of Gas to be Measured

The spectrum of the gas to be measured, as obtained from the interferogram, can be stored in the form of a transmission or an absorption spectrum. The original spectrum obtained from the interferogram is called a raw spectrum, as it contains other factors in addition to the information coming from the gas to be measured. These factors include the instrumental effects connected with the background spectrum, as mentioned earlier. The ratio of the raw spectrum and the background spectrum is calculated by the instrument's computer, yielding the spectrum of the sample gas. As an example, the transmission spectrum of polystyrene is presented in Fig. 6.13. Absorption appears as peaks directed downwards. It also appears from the figure that polystyrene is very transparent to a wide range of wavenumbers (wavelengths). Similarly, the absorption

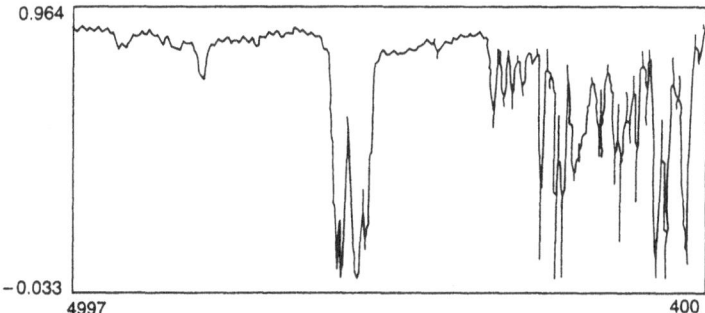

Figure 6.13. FTIR transmission spectrum of polystyrene [112].

spectrum can be calculated based on the measurement of the interferogram as performed by the spectrometer. Such a spectrum for polystyrene is presented in Fig. 6.14, which can be compared with the transmission spectrum (Fig. 6.13). Peaks in these spectra appear at the same positions, but have opposite directions. An interferogram registered from polystyrene is presented in Fig. 6.15.

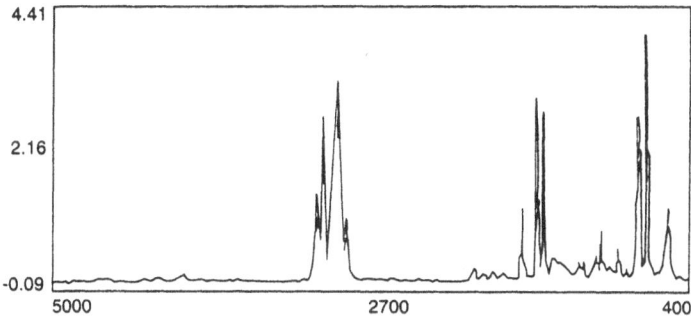

Figure 6.14. FTIR absorption spectrum of polystyrene [112].

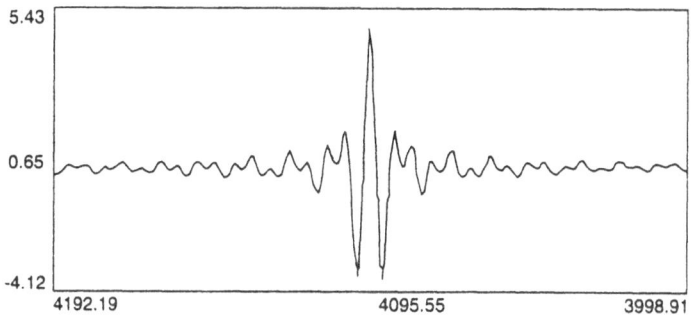

Figure 6.15. Interferogram registered from polystyrene [112].

Elimination of Interference from Unwanted Gases

An essential part of FTIR gas analysis is the elimination of the intereference caused by gases other than the gas to be measured. This is done mathematically and is illustrated in Figs 6.16 and 6.17. The uppermost spectrum of Fig. 6.16 has been registered above the flame of powdered peat, where there are no hydrocarbons in the flue gas [105]. The absorption peaks of the oxides of nitrogen appear pure in this spectrum. On the other hand, in lower spectra, which have been registered inside the flame, the absorptions caused by hydrocarbons are clearly seen. The absorption band of methane overlaps with the weaker absorption peak of N_2O ($1300\,cm^{-1}$), and the strongest band of N_2O remains covered by CO_2.

By means of the spectrum analysis, the problem caused by the overlapping absorption bands can be eliminated to certain extent. In the case illustrated in Fig. 6.16, the concentration of methane can be determined based on the absorption peak in the region, where no interference occurs ($3000\,cm^{-1}$), and the peak overlapping with the peak of N_2O can be removed from the spectrum mathematically. Thus the pure N_2O absorption remains. This is shown in Fig. 6.17, where also the strong absorption peak of CO_2 at $2300\,cm^{-1}$ has been removed. This leaves the other absorption peak of N_2O uncovered. In a similar way, the covering effect of water vapour has been reduced from the spectrum.

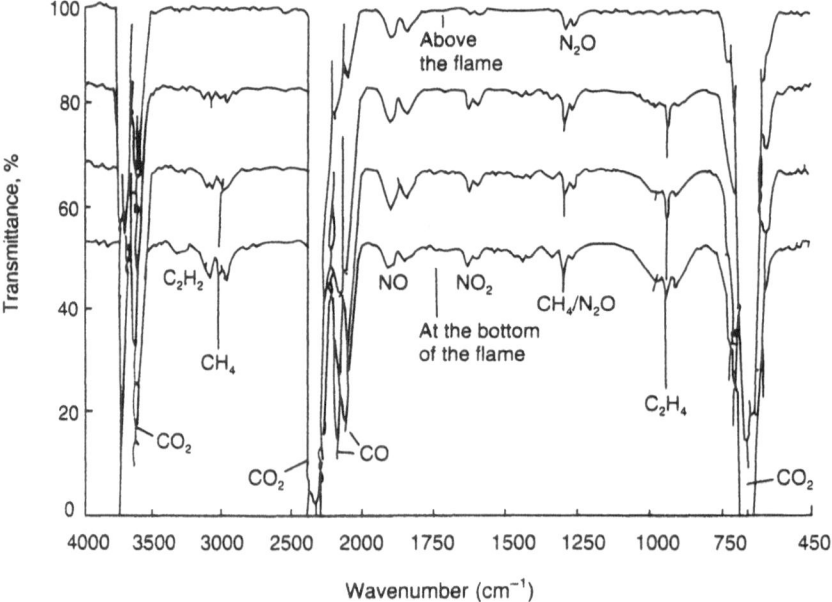

Figure 6.16. Absorption spectra of gases in different parts of the flame of peat powder. The uppermost spectrum has been registered from above the flame, and the lower spectra, in order, from the inner parts of the flame [105].

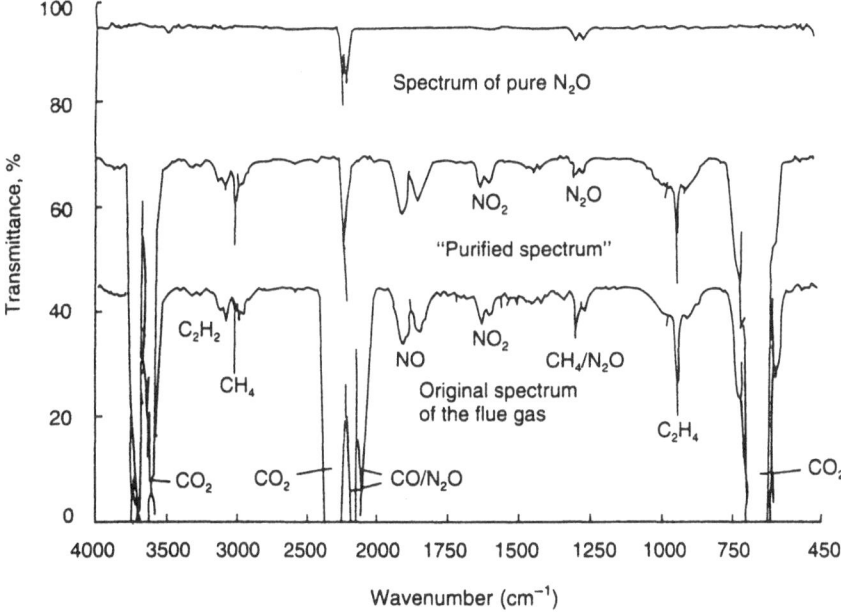

Figure 6.17. "Purification" of the spectrum from interfering compounds mathematically [105].

In research on the application of FTIR technology to the emission gas analysis, the lowest concentration detectable with a specific instrument and arrangement was estimated [105]. The estimated values are presented in Table 6.7. The sensitivity of the FTIR system can be improved by a factor of a hundred or so by using a liquid-nitrogen cooled MCT detector and by making the absorption distance longer.

Table 6.7. Lowest detectable concentrations of different gas components in emission gases (DTGS detector, absorption distance 6.75 m, resolution 4 cm^{-1}, one scan per spectrum) [105]

Compound	Wavenumber range (cm^{-1})	Detection limit (vppm)[a]
C_2H_2	3177–3390	2
C_2H_4	808–1152	3
C_2H_6	2830–3085	2
CH_4	1210–1380	5
CO	2087–2181	10
CO_2	3540–3646	10
SO_2	1320–1390	1
NO	1780–1964	5
NO_2	1567–1666	0.3
N_2O	1237–1330	1

[a] Parts per million by volume

Measurement of Concentration

The measurement of concentration is based on the use of calibration gases. These gases can be bought as prepared to appropriate concentrations comparable to concentrations to be measured, or they can be prepared by mixing. Spectra obtained from the calibration gases are stored in the memory units of the automatic FTIR instruments.

The measurement of concentration is naturally based on the strength of the absorption. The integrated intensity of a certain absorption peak is proportional to the concentration of the gas component in the gas mixture.

Calibration

The same measurement chamber should be used for both calibration and measurement [106]. The calculation of correction factors can thus be avoided, which would otherwise be necessary, as the absorption is enhanced with increasing optical length. It is important to remove solid materials from both the calibration gas and from the gas to be measured, as solid particles can scatter or bend infrared radiation so that the radiation passing through the chamber does not correspond to the true transmittance. The pressure and temperature of the calibration gas and the gas to be measured should also be maintained at the same value. When measuring hot emission gases, for instance, the absorption bands can differ significantly from the bands of the calibration gas. At elevated temperatures, the intensity of the peak absorption is decreased while the width of the band is increased.

The concentration of the calibration gas should be of the same order of magnitude as the concentration of the gas to be measured. If the calibration spectra available for the measurement programs of the spectrometers have been registered from calibration gases of different concentrations, the extrapolation made by these programs can lead to inaccurate results.

A certain wavenumber range is restricted from the calibration infrared spectra applicable to the gases studied, and this is fed to the computer of the spectrometer.

6.4.7. Resolution in the Measurement of the Concentration of Gases

The resolution used in the gas measurements is normally $1-8\,\mathrm{cm}^{-1}$. The most generally used value is $4\,\mathrm{cm}^{-1}$, which is sufficient for ordinary measurements. The selection of the resolution is, of course, dependent on how precise a knowledge of the fine structure of the spectrum is needed. When selecting the resolution, significant factors include the following:

- Does the result of the measurement depend on the resolution?
- Does the detection limit change when the resolution is changed?
- Which resolution is required for the measurement of the concentration of a certain gas at a certain concentration level?

Use of the highest resolution ensures the most efficient separation of spectra from each other. The analysis time and the capacity requirements for the storage of the information increase strongly, however, as the resolution is improved, and the signal-to-noise ratio can decrease. To balance these dual requirements, an optimum resolution should be found.

In cases where absorption bands are wide compared with the resolution selected, high-resolution measurement can produce high noise peaks compared with the peaks caused by absorption bands. The signal-to-noise ratio as well as the detection limit can thus be impaired [113, 114]. It is then practical to use lower resolution. When the width of the band is of the same order of magnitude as the resolution, increasing the resolution can lead to a significant increase in both the absorption peak and the noise peak. In such cases the influence of the change of the resolution cannot be readily predicted, and an assessment of the resolution requirement has to be made in each case.

The situation is still more complicated if there are sharp peaks of water vapour in the measuring range. Then a higher resolution can point out more clearly the influence of the peaks of water vapour, and show proper preconditions for carrying on the measurement. One way could then be to utilise some peaks of water vapour for the elimination of the influence of water vapour and to use lower resolution after.

In Fig. 6.18 there is an example of a compromise between the resolution and the signal-to-noise ratio. In this case the width of the absorption band is of the same order of magnitude as the resolution. When the resolution is changed from $0.1\,cm^{-1}$ to $0.5\,cm^{-1}$, the peak indicating the absorption of the band is reduced to about a half, and the relative noise is reduced a little more. The selection of the best resolution for each case is thus not quite straightforward. The solution is made still more difficult by the fact that there are several peaks of water vapour and carbon dioxide in the wavenumber range $670\text{--}720\,cm^{-1}$. The use of a higher resolution can sometimes appear to be necessary, so as to be able to separate the absorption peaks of benzene from the peaks of water vapour and carbon dioxide.

Finding the practical solution is easier in the case shown by Fig. 6.19, where the absorption band of 1,1,1-trichlorethane, as measured from air, is wide compared with the resolution. Changing the resolution from $0.1\,cm^{-1}$ to $0.5\,cm^{-1}$ has only a minor effect on the absorption peak, but reduces the noise crucially. From the figure it becomes clear that in this case, the lower resolution leads to a better detection limit.

In some investigations, resolution requirements for the measurement of different gases has been estimated [114–116]. Some values are presented in Table 6.8. The measuring ranges have been selected so as to minimise the interference effects caused by atmospheric gases and vapours.

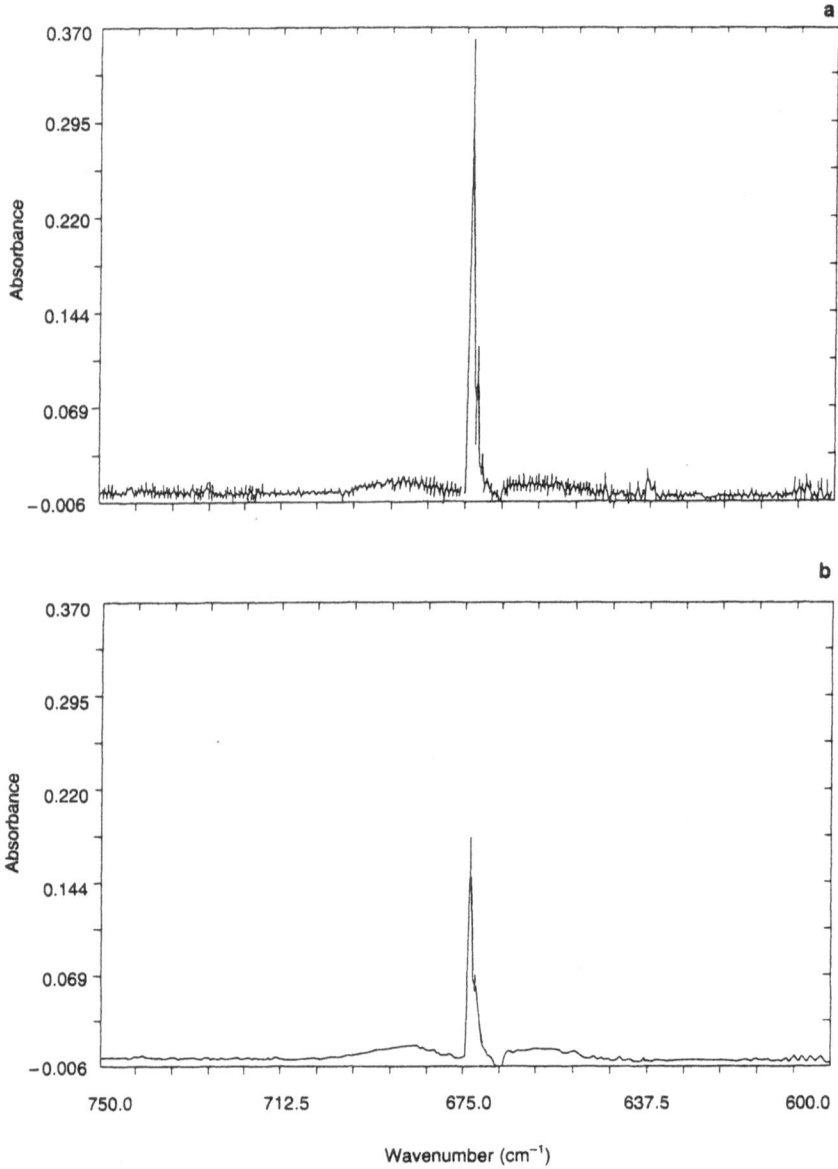

Figure 6.18. Absorption spectra of benzene. The resolution is of **a** $0.1\,\text{cm}^{-1}$ and **b** $0.5\,\text{cm}^{-1}$ [113].

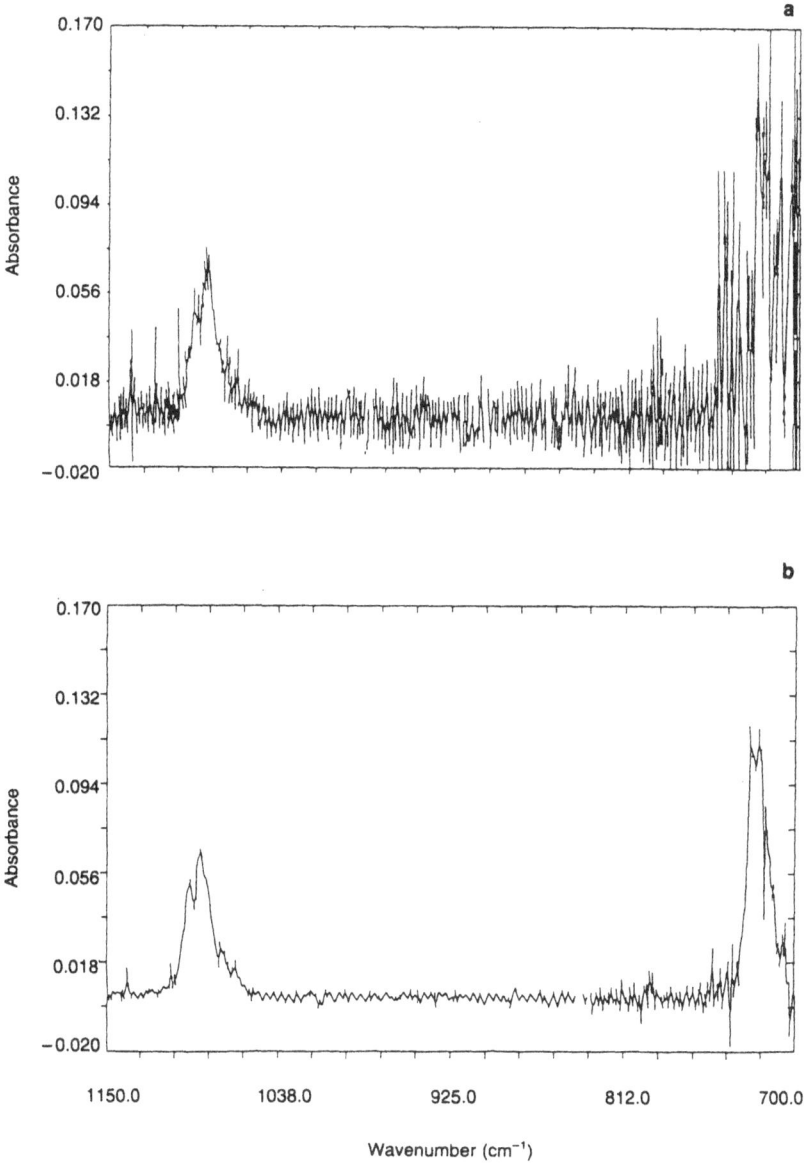

Figure 6.19. Absorption spectra of 1,1,1-trichlorethane as measured from air. Open measuring path is 117 m and the resolution **a** 0.1 cm^{-1} and **b** 0.5 cm^{-1} [113].

Table 6.8. Practical ranges of wavenumbers for the FTIR measurement of some gases and the lowest practical resolution of the measurement [116]

Compound	Wavenumber range (cm^{-1})	Lowest resolution (cm^{-1})
Acetone	1287–1167	8
Arsine	3132–2106	2
Diborane	2522–2515	0.5
o-dichlorobenzene	1060–1002	8
2-ethoxyethanol	1192–984	8
Freon[R] 11	876–813	8
Freon 13B1	1137–1031	8
Freon 22	1193–1063	8
Nitrogen trifluoride	960–833	8
Phosphine	2440–2390	4
Sulphur hexafluoride	965–915	8

6.4.8. Influence of Some Factors on Measurement of Gas Concentrations

Temperature

The rotational and vibrational energy states of gas molecules follow the Boltzmann distribution law, which is dependent on temperature [106]. At elevated temperatures, the relative proportion of the high energy levels of the molecules is increased and the boundaries of the vibrational–rotational bands are changed. At high temperatures, hot bands appear in the spectrum, and these affect the average distances of the absorption spectrum bands. In the band models, the temperature dependence is taken into account by determining the model parameters as a function of temperature separately for each wavenumber.

Pressure

The fine structure of a gas spectrum can be clarified in some cases by reducing the total pressure of the gas to be measured. The influence of the pressure on the spectra of ethane and propane is presented in Fig. 6.20. The fine structure observed in the spectrum of ethane is clearly increased, as the total pressure is decreased from 760 torr to 10 torr. In the fine structure of propane, instead, no clear change can be observed in this measurement performed with the resolution of 0.125 cm^{-1}. There are so many spectral lines near one another in the spectra of heavy and complicated molecules, that a very low pressure would be needed to separate them.

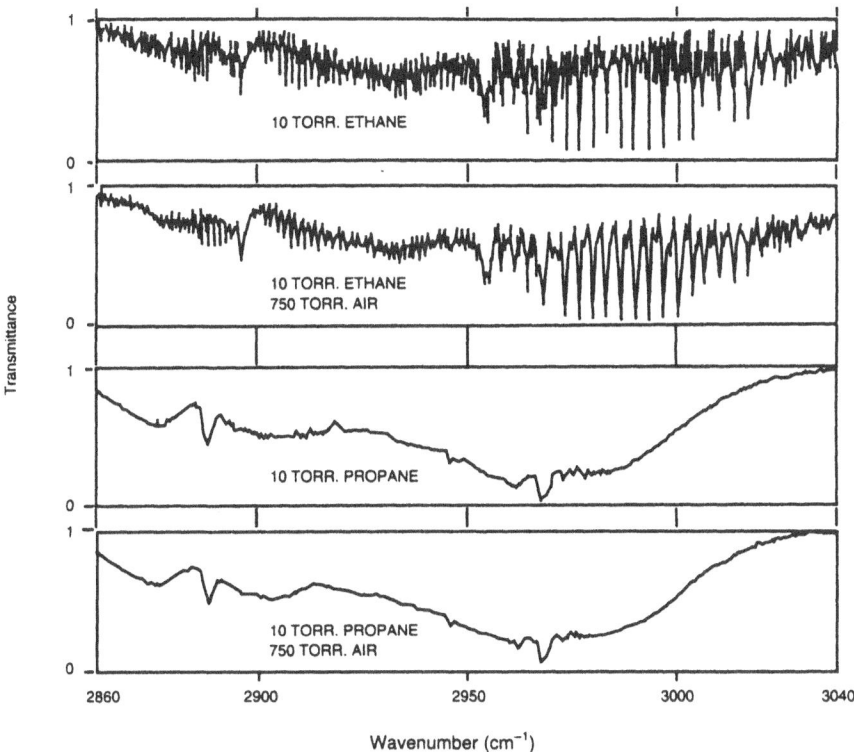

Figure 6.20. Effect of pressure on the fine structure of the infrared spectra of ethane and propane [115].

Nonlinearity

Sometimes the intensity of an absorption peak does not follow regularly the change in the concentration of the gas. This can be due to some substances present in the measuring chamber, either as a result of reactions taking place in the chamber or coming in for some other reason. In Fig. 6.21, an example is given concerning a spectrum of diborane. The absorption peak on the left does not follow the concentration, which must be determined using the peak on the right.

Overlap of Absorption Bands when Measuring Emission Gases

The following consideration is based on the estimates made in an M.Sc. thesis [106]. Every compound must be examined separately and the most appropriate absorption band must be chosen. This band is not always the strongest one.

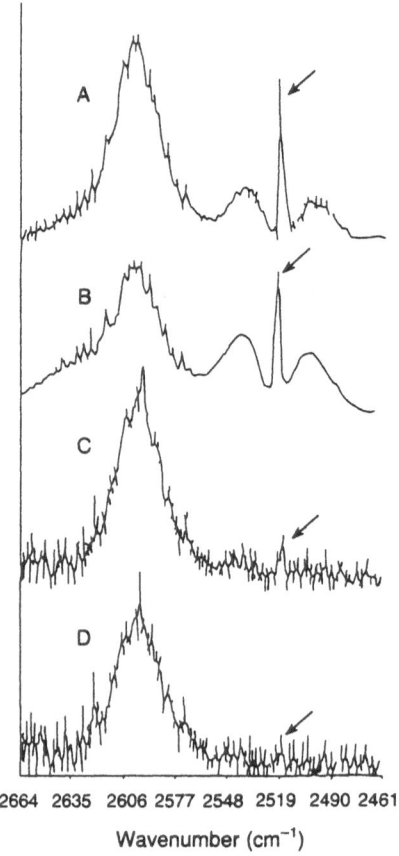

2664 2635 2606 2577 2548 2519 2490 2461

Wavenumber (cm⁻¹)

Figure 6.21. Infrared spectrum of diborane having a nonlinear peak at the left [116]. The concentration of diborane is decreased by the factor of ten at each stage from A (23 ppm) to D (0.023 ppm).

For example, the site of the strongest absorption band of C_2H_2 is 729 cm^{-1}, but it is covered strongly by the band of CO_2, especially in flue gas measurements. The absorption band at 3287 cm^{-1} can often be used, as the absorption bands of other compounds do not disturb this band. A very high concentration of water vapour can disturb the examination of C_2H_2.

For the examination of C_2H_4, its strongest line at 949 cm^{-1} is suitable, because there are no interfering compounds near in the spectrum. There is, however, a strong peak of NH_3 in the same range, but this compound does not often occur in, e.g. flue gases. Nevertheless, these two gases can be well separated from each other. High concentrations of carbon dioxide can cause interference.

C_2H_6 can be examined in the range 3085–2830 cm^{-1}, which consists partly of its strongest band and of several weaker bands. The strongest absorption band is partly overlapped by the bands of C_2H_4 and CH_4.

The absorption band of CH_4 mentioned above is not suitable for the determination of the concentration, as it is covered strongly by C_2H_6 and C_2H_4. Another strong absorption band of CH_4 is in the range of 1380–1210 cm^{-1}. A band of C_2H_2 also occurs in this range, but it is not very strong and it interferes only when the concentration of C_2H_2 is high. The occurrence of water vapour in the spectrum disturbs the examination of the above-mentioned absorption band of CH_4.

The situation of the only strong absorption band of CO is 2143 cm^{-1}. It is by the side of a strong band of CO_2, and there is one medium-strong band of CO_2 partly overlapping the CO band.

There are several bands available for the examination of CO_2, and their ranges of wavenumbers are 3646–3540 and 700–580 cm^{-1}. Abundant water vapour can cause interference with the former band.

The situation of the absorption band of NO is 1876 cm^{-1}. This is in the range of a strong absorption of water vapour. The band is partly overlapped by an absorption band of C_2H_4 and a high concentration of C_2H_4 causes interference when examining NO.

The absorption band of NO_2 is situated in the middle of a strong absorption region of water vapour. After the elimination of water vapour, the concentrations of NO_2 can be determined reliably, as there are no additional compounds in the same region.

In addition to water vapour, CO_2 disturbs the examination of several compounds. By removing CO_2 from flue gas, the detection limit of CO can be improved, as the calibration range can be extended over the whole region of the band of CO. Furthermore, the strongest band of C_2H_2 can be taken into consideration, if CO_2 can be removed. The strongest band of HCN, for instance, is also covered by CO_2.

6.4.9. Multicomponent Analysis of Gases Based on FTIR spectra

In a multicomponent analysis we have the infrared spectrum of an unknown mixture of gases, for which we have no knowledge of the constituent gases, not to mention their partial pressures [117]. Multicomponent analysis developed for the FTIR techniques can be used to determine the partial pressures of the system. It has also the advantage of being able to reveal a gas component, which could not be anticipated to be present.

The multicomponent analysis is based on a large set of library spectra of pure molecular gases, measured at known pressures and with the same interferometer as the unknown sample [117]. By using these pure spectra it is possible to calculate the partial pressures of the pure gases in the mixture, with error limits. The errors in the values obtained arise from the measurement noise in the spectra. The analysis is based on the calculation of such partial pressures that best explain the mixture spectrum.

The multicomponent analysis method is demonsrated in Fig. 6.22. The coefficients are calculated based on the vector calculation developed [117] using

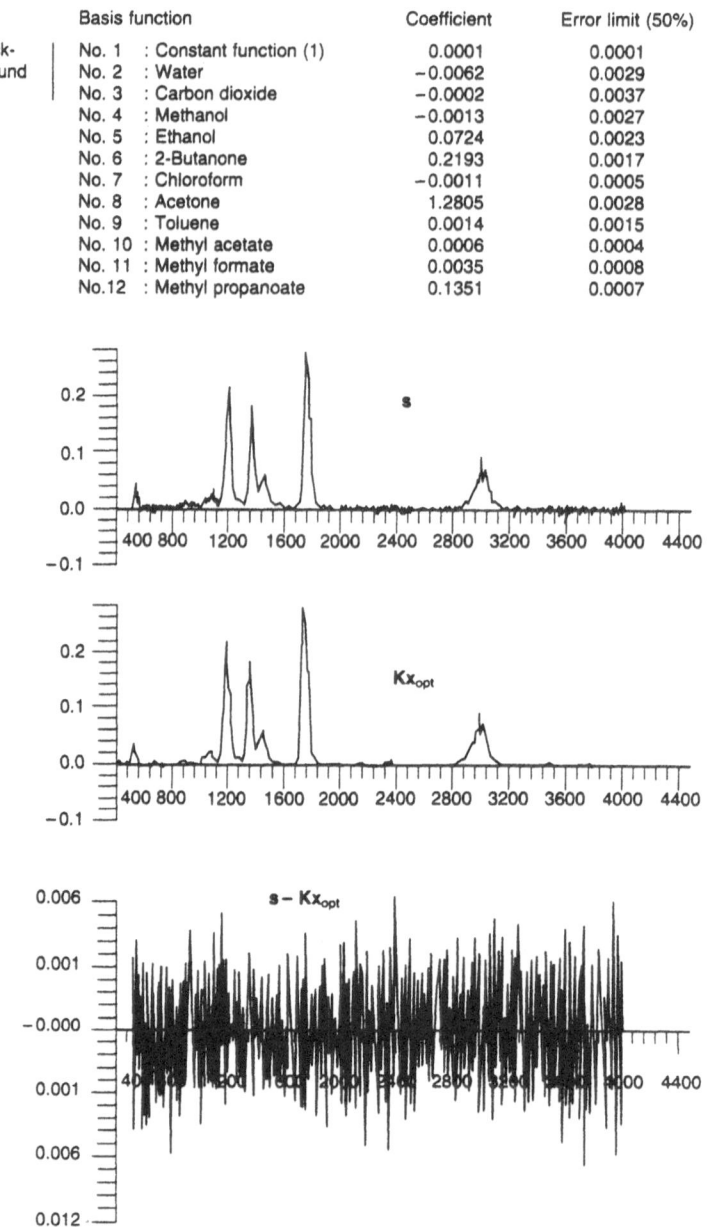

	Basis function	Coefficient	Error limit (50%)
back- ground	No. 1 : Constant function (1)	0.0001	0.0001
	No. 2 : Water	−0.0062	0.0029
	No. 3 : Carbon dioxide	−0.0002	0.0037
	No. 4 : Methanol	−0.0013	0.0027
	No. 5 : Ethanol	0.0724	0.0023
	No. 6 : 2-Butanone	0.2193	0.0017
	No. 7 : Chloroform	−0.0011	0.0005
	No. 8 : Acetone	1.2805	0.0028
	No. 9 : Toluene	0.0014	0.0015
	No. 10 : Methyl acetate	0.0006	0.0004
	No. 11 : Methyl formate	0.0035	0.0008
	No.12 : Methyl propanoate	0.1351	0.0007

Figure 6.22. A mixture spectrum s, the linear combination Kx_{opt} of the library spectra that best explains s, and the residual spectrum $s − Kx_{opt}$, which then remains unexplained (appears magnified in the figure) [117]. In the table at the top, the components x_{opt} with error limits are shown. The three first components are the background spectra. These are all the spectra or functions, which are present also in the background measurement or which can arise from error data in the interferogram. They can have negative coefficients as well as positive. The mixture spectrum s and the library spectra K^i are measured with identical arrangements, but extra artificial noise has been added to s in order to simulate poorer accuracy.

a linear combination $\mathbf{Kx}$ of the pure spectra that is as near as possible to the vector $\mathbf{s}$ denoting the measured mixture spectrum. Here the 50% error limits are used, because they give a good idea of the orders of magnitude of the errors. In this analysis a rather low number of 12 library spectra are used. In the figure we see mixture spectrum $\mathbf{s}$ to be analysed by the set of 12 library spectra. In addition, we see the linear combination $\mathbf{Kx}_{opt}$ of these library spectra that best explains $\mathbf{s}$. Also shown is the remainder spectrum $\mathbf{s} - \mathbf{Kx}_{opt}$, which then remains unexplained. As can be seen, the remainder spectrum consists of pure white noise. This indicates that the analysis has been succesful and that the 12 library spectra are sufficient to explain the measurement. At the top of Fig. 6.22 we can see the result of the analysis of the optimal coefficient vector $\mathbf{x}_{opt}$. The partial pressure of a library gas j is now obtained by multiplying its measuring pressure by its coefficient $x_{opt,j}$. Those components, which do not exist in the mixture, have small positive or negative coefficients with the same order of magnitude as their error limits.

Optimal Resolution

For the analysis, the wavenumber range is fixed. The sampling interval in the interferogram is also fixed according to the Nyqvist sampling theorem [117]. It still remains to choose the length of the registered interferogram or the amplitude of the mirror movement, which in turn is determined by the number of data used, N (single-sided interferogram). The corresponding number used by the fast Fourier transform (FFT) algorithm is then $2N$.

The error limits are directly proportional to the standard deviation of the spectral noise and inversely proportional to the square root of the number of the data N. When the number of the data is diminished by some factor $1/k$, the coefficient $1/\sqrt{N}$ is increased by factor $k^{1/2}$. It appears, however, that the standard deviation of the noise is diminished by factor $k^{-1/2}$ so that these two factors cancel the influence of each other when reducing the resolution.

In addition to the factor containing the standard deviation of the spectral noise divided by $\sqrt{N}$, the expression for the error limits still contains another factor, which can be the source of changes in error limits when changing the resolution [117]. It can be pointed out that this factor is not an explicit function of N, but rather depends only on the shapes of the library spectra, when the number of library spectra M is fixed. All linear changes, where all the library spectra are multiplied by some constant coefficient C, change this factor by constant C^{-1}.

In practice, the spectra are always computed from the corresponding interferograms by application of the FFT algorithm [117]. A fundamental property of this algorithm is that the data interval in the spectrum is $1/(2N\Delta x)$, where Δx is the sampling interval in the interferogram. So, when the number of the data is diminished by factor $1/k$, the data interval in the spectral domain is increased by factor k. As long as the data interval ($\approx$ resolution/1.21) stays smaller than the full width at half-height (FWHH) of the spectral lines, there exists at least one datum at every line, and the shape of the spectral lines does not vary considerably.

This means that there is only very little benefit to be gained from employing a resolution better than the width of the spectral lines. In the interferogram domain this means that the interferogram can be truncated.

Some technical advantages can be gained by reducing the resolution [117]. The signal can be increased by increasing the radius of the radiation source. The amount of computation can also be decreased. When the number of data is decreased by factor $1/k$ as described earlier, we are able to register k interferograms in the same amount of time previously used to register one. Because Fourier transformation is a linear operation, co-adding these interferograms means that the corresponding spectra are also co-added. The errorless spectra **e** remain the same in every measurement, which means that they become multiplied by k in the summation. This means a simple linear change of the spectra, which in turn means that the error limits become multiplied by k^{-1} [117]. The noise of **s**, on the other hand, is different every time, and it does not sum linearly. The standard deviation of the noise is increased by factor $k^{1/2}$. The total effect is that the error limits become multiplied by factor $k^{-1/2}$.

When all the different effects mentioned above are gathered together, it becomes apparent that if we can freely change all the parameters of the interferometer, as small a resolution as possible should be used [117]. The number of data N should, however, be at least two or three times as large as the (maximal) number of the library spectra for the structure of the residual spectrum **s** − **Kx** to be examinable.

6.5. Measurement of Gas Concentrations by Ultraviolet, Visible and Infrared Light

In addition to the infrared light region, absorption peaks at the visible and ultraviolet regions can be used in the measurement of gas concentrations. Several gases can be measured based on the absorption in these regions. Figure 6.23 shows absorption maxima caused by some gases at different wavelengths.

In a gas analyser using these wavelength bands, there are two flow-through analysis cells (3, Fig. 6.24), both of which contain the measurement chamber (6) and the reference chamber (8) [87]. The gas to be analysed is conducted into one of the cells, depending on whether the wavelength of the strong absorption of the gas in question is in the infrared, visible or ultraviolet region of the spectrum. The light sources associated with the cells produce either infrared (1) or visible and ultraviolet (2) light. The absorption of the light passing through

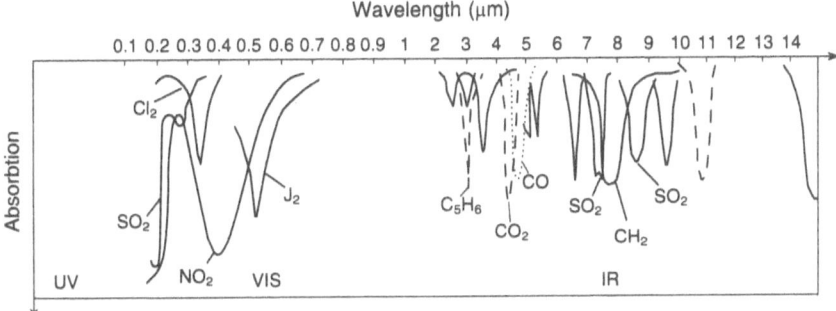

Figure 6.23. Absorption maxima caused by some gases in the wavelength region of ultraviolet, visible and infrared light [87].

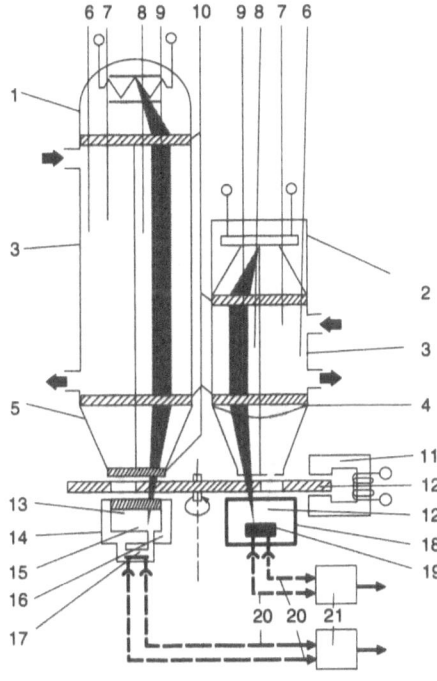

Figure 6.24. Instrument for the measurement of gas concentrations based on the absorption of ultraviolet, visible and infrared light [87]. 1, infrared light source and reflector; 2, visible and ultraviolet light source and reflector; 3, analysis cell; 4, focusing lens; 5, filter cell; 6, measurement chamber; 7, measurement beam; 8, reference chamber; 9, reference beam; 10, window; 11, eddy-current drive; 12, light chopper wheel; 13, transmitted light pulses (intensity dependent upon concentration); 14, pneumatic detector (IR); 15, detector main chamber; 16, expansion chamber; 17, microflow sensor; 18, photodetector (VIS/UV); 19, photocell; 20, signal pulses; 21, signal processing electronics.

the measurement chamber (7) then depends on the concentration of the gas measured in the measurement chamber. The rest of the light (9) passes through the reference chamber, which usually contains nitrogen.

The light pulses transmitted alternately through the measurement and reference chambers are received by the detectors below the sample cells. The pulsing is produced by holes situated in a chopper wheel (12) rotating between the detectors and the analysis cells. The difference in the intensities of the alternate pulses is converted into the measurement signal.

The detector for the infrared light is a gas-specific, pneumatic detector (14). It consists of two chambers. In the gas-filled main (15) and expansion (16) chambers, the two light pulses of different intensities from the measurement and reference chambers are transformed, due to different heating effects, into an oscillatory gas flow between the two chambers. This oscillatory flow is converted into an electrical signal (20) by a microflow sensor.

The photodetector (18) used for the visible and ultraviolet light converts the different light intensities from the measurement and reference chambers into a corresponding electrical signal.

6.5.1. Differential Optical Absorption Spectroscopy (DOAS)

Most of the gases to be measured from emissions have an appropriate infrared spectrum. Some difficulties are encountered, however, with a few gases. These difficulties arise for the following reasons [118]:

- There is weak absorption in the infrared region.
- The spectrum of the compound appears in the same region as that of water vapour or carbon dioxide.
- The infrared absorption spectrum does not exist (e.g. O_2 and other molecules formed by two similar atoms).

Several compounds, the detection of which is hampered for these reasons, can be measured using absorption in the region of ultraviolet (UV) or visible light. This can be done by DOAS techniques, in which the absorption is measured at many (e.g. 1000) discrete wavelengths in the ultraviolet and visible region. The dispersion of light is done using a grating spectrometer. High-pressure xenon lamps can be used as light sources [119]. These lamps produce a light spectrum ranging from about 200 nm upwards to the wavelengths of the visible light. The measurement principle of a DOAS spectrometer is presented in Fig. 6.25. In emission measurements, the DOAS techniques can be applied to measure across the emission gas duct or across the smoke plume after the pipe-end.

In DOAS analysis, molecular concentrations are determined based on the Lambert–Beer law. The calculation of the concentration, carried out by the computer in the central unit of the system, is based on library spectra recorded under controlled laboratory conditions in the computer memory at 1000 different wavelengths [119]. The calculation gives the concentration as the average value of the results obtained from those 1000 wavelengths. The standard deviation is also calculated to give the degree of reliability.

One DOAS system uses an optic fibre to conduct the light from the receptor unit to the central unit [119]. The measurement result is an average concentration of the gas component measured between the emitter and receptor units.

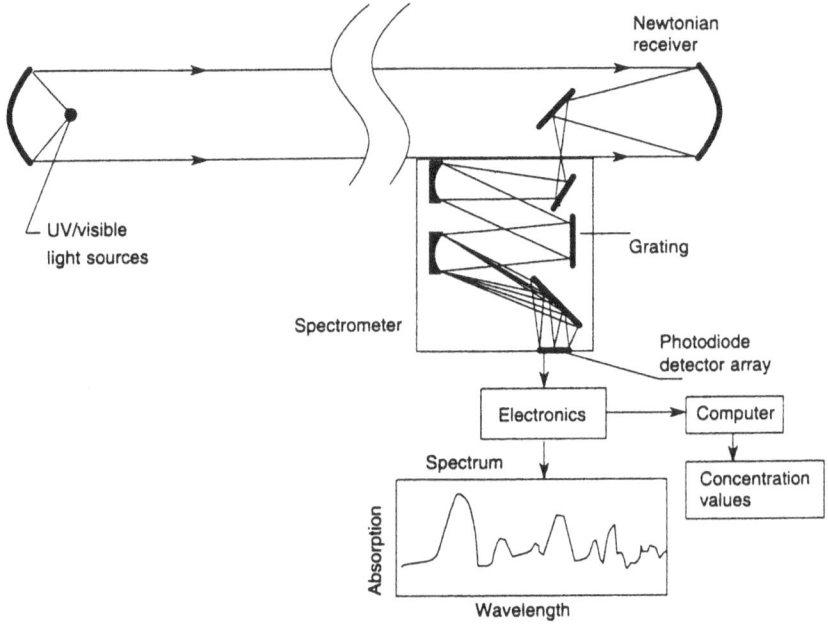

Figure 6.25. Block diagram of a differential optical absorption spectrometer (DOAS) in long-path operation, showing the key optical elements and the resulting spectrum [120].

6.6. Features of FTIR Techniques with Reference to Other Methods

The infrared region is the most important spectral region for sample identification [79]. The infrared method makes it possible to identify and quantify all molecules except diatomic, homonuclear molecules. With a single analyser and a single measurement several compounds can be identified and their concentrations determined from ppb levels upwards with results available in seconds.

FTIR techniques can be used for both emission measurements and air quality measurements. Multireflection cells are often used in emission measurements. These make the optical path several metres long and ensure high sensitivity in emission measurements.

The operation of FTIR analysers is practically continuous. This is an advantage as compared with methods based on sampling and laboratory analysis. Concentrations of hydrocarbons, for instance, can be determined separately by the FTIR analysis, whereas the FID techniques can measure methane and non-methane hydrocarbons in addition to the total concentration of hydrocarbons, as expressed in equivalent values. FTIR techniques are also applicable to the real-time measurement of the reduced organic sulphur compounds, which characterise the emissions from pulp industry. These measurements normally require gas chromatographic analysis. Due to weak absorption, however, hydrogen sulphide (H_2S) at low concentrations cannot be easily measured by infrared methods. Other common emission gases, such as sulphur and nitrogen oxides as well as carbon oxides, can be analysed by FTIR techniques. For example, the concentration of nitrous oxide (N_2O) can be followed using FTIR techniques. There are still very many other gases, for which the FTIR analysis is useful. This gives an idea about the capacity of an appropriate FTIR analyser for emission measurements, where separate analysers are otherwise needed for practically all gas components analysed. This is the case especially when modern multicomponent analysis is used.

One of the major environmental problems of today is determining the origin, identity, and amounts of volatile organic compounds (VOCs) [113]. There are many potential sources of VOCs, e.g. chemical dump sites, landfills, grain elevators (where fumigants have been used), industrial sites, surface waters, surface impoundments, container storage areas, chemical spills, leaking tanks, land treatment units, waste piles, water sludge deposits, air stripping plants, and indoor areas. VOCs can be measured rapidly and reliably by FTIR analysers. In open air measurements, for instance, long open paths of possibly hundreds of metres are used. This increases the sensitivity and decreases the detection limit of gas components. This technique also results in an average concentration value over the path. Sampling methods, including continuous monitoring of sulphur dioxide and nitrogen oxides, give the concentration value for one point.

Recent development of FTIR analysers has created instruments that can be applied to the measurement of emissions at their sources [79]. Both constructional features and mathematical operations have been developed towards better usability and field operation. These instruments operate practically continuously and can record concentrations of several gas components from emission gases almost simultaneously. FTIR analysers are versatile instruments and are increasingly applied to practical process emission measurements.

Chapter 7

Other Measurement Technologies

7.1. Electrochemical Measuring Cells

By the use of electrochemical measuring cells, it has become possible to construct portable measuring instruments which can simultaneously record concentrations of several gas components. In the instrument, there is a specific cell for each gas to be measured. Electrochemical cells have been produced to measure the concentrations of oxygen, carbon dioxide, carbon monoxide, hydrogen sulphide, nitrogen oxides and sulphur oxides from emission gases.

The operation of an electrochemical cell is based on an oxidation–reduction reaction, which is specifically characteristic of the gas to be measured. In the following, the measurement of the concentrations of oxygen and carbon monoxide is considered as an example.

7.1.1. Oxygen Cell

At the cathode of the measuring cell, oxygen is reduced to hydroxyl ions, which move through the electrolyte to the anode, and oxidise the metallic anode. The current generated is proportional to the velocity of the consumption of oxygen according to the Faraday law. Via this reaction all oxygen coming to the cathode is consumed, so that the concentration of oxygen at the cathode remains practically zero. The sensor cell operates on limiting current basis, and the changes in the cell voltage do not affect the current (Figs 7.1 and 7.2).

The arrival of oxygen at the cathode is limited by diffusion. Oxygen can only get to the cathode through a diffusion barrier. This barrier can be a plastic film or a gas barrier in a capillary tube. Due to the diffusion barrier, the rate at which oxygen arrives at the cathode is dependent on the concentration of oxygen in the gas to be measured. The current of the sensor cell is, in turn, dependent on the the rate oxygen arrives, as was noticed above. Altogether, the cell current is consequently dependent on the concentration of oxygen in the gas to be measured.

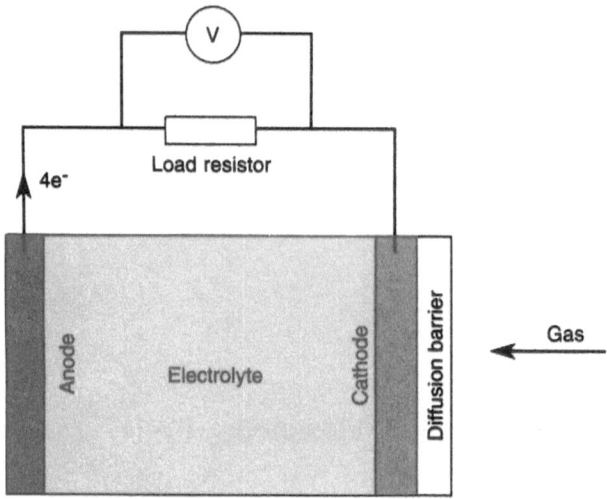

Figure 7.1. Scheme of an electrochemical oxygen cell. The reactions are

Anode: $2\,Pb + 4\,OH^- \longrightarrow 2\,Pb(OH)_2 + 4\,e^-$

Cathode: $O_2 + 2\,H_2O + 4\,e^- \longrightarrow 4\,OH^-$

Cell: $2\,Pb + O_2 + 2\,H_2O \longrightarrow 2\,Pb(OH)_2$

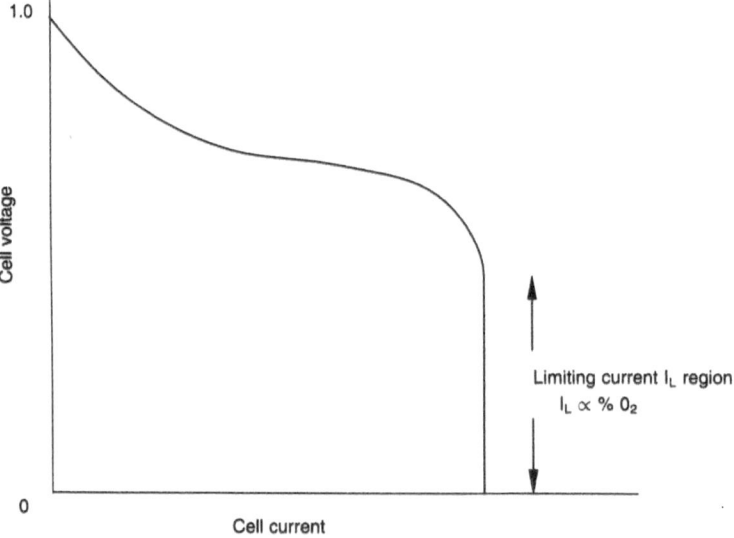

Figure 7.2. Characteristic current–voltage curve of an electrochemical oxygen cell. The region of the limiting current is the vertical part of the curve.

7.1.2. Carbon Monoxide Cell

The reactions in the measurement of carbon monoxide are

$$CO + H_2O \longrightarrow CO_2 + 2H^+ + 2e^- \quad \text{(anode reaction)}$$

$$\tfrac{1}{2}O_2 + 2H^+ + 2e^- \longrightarrow H_2O \quad \text{(cathode reaction)}$$

In this case also, the flow rate of carbon monoxide coming to the indicator electrode is controlled by diffusion. Oxygen needed at the cathode comes from ambient air, again controlled by diffusion.

When using capillary diffusion barriers, the dependence of the measurement signal on temperature and pressure remains low.

7.1.3. Cells for Toxic Gases

Electrochemical cells are available for the measurement of CO, SO_2, NO, NO_2, H_2S, Cl_2 and HCl. The three electrode cells used for toxic gas measurements are micro fuel cells [121]. The use of a gaseous diffusion barrier results in a direct response to volume concentrations.

The three-electrode toxic gas cell consists of a sensing electrode and a reference electrode separated by a thin layer of electrolyte. The gaseous diffusion barrier ensures that the flow of gas to the sensing electrode does not exceed its electrochemical activity. Gas diffused to the sensing electrode reacts at the surface of the electrode, either by oxidation (e.g. CO, H_2S, SO_2, H_2, HCN, HCl), or by reduction (NO_2, Cl_2). The specific electrode material used determines which gas the cell will react to. The counter electrode built into the cell acts to balance out the reaction at the sensing electrode by reducing oxygen in air to water [121].

7.1.4. Methods of Use

Ambient temperature electrochemical cells cannot be inserted directly into a flue gas stream [121]. It is necessary to introduce a sampling system which cools the flue gases into the measurement scheme. Hence the major use for these cells as emission gas monitors has been to incorporate them into portable instruments. There are many significant points that instruments incorporating this type of electrochemical cells should have included in their design, such as:

- The gas sample system should be designed with appropriate filters to remove highly corrosive constituents and particulate material.
- Sample gas lines should be in materials which do not absorb the gas components. This is particularly important for gases such as SO_2 and NO_2 where plastic materials (PTFE, polytetrafluoroethylene) is normally recommended.

- Condensing conditions at the cell must be avoided.
- If used intermittently, cell life is improved by purging toxic gas cells after use.

7.1.5. High Temperature Sensors

Development of electrochemical sensors has produced cells which can be used in *in situ* probes, giving the advantages of this measuring principle. It is possible to monitor SO_2 and NO_x continuously with electrochemical cells directly inserted into flue gas streams up to 260°C [121].

7.2. Solid State Gas Sensors

The development of solid state gas sensors is aimed at improving and simplifying gas concentration measurements. Sensors have been developed for the need to measure the concentrations of both emission gases and impurity gases in the air environment.

The main interest was originally directed towards ceramic metal oxides, the electrical properties of which have been noticed to change when gases react at their surfaces. These electroceramic materials are also known to resist severe conditions. Later on, organic semiconductors were also the object of research work.

Solid state sensors have already been studied and developed for quite a long time. The basis for the development has been those advantages which they are expected to offer compared with the methods available. When developing continuously indicating gas sensors, attention is paid to properties such as

- Sensitivity and short response time
- Resistance to aggressive conditions
- Simple and convenient operation
- Applicability to the measurements of both process emission gases and air impurity gases
- Practical production and low price

The operation of electroceramic gas sensors is based on the dependence of some electrical characteristic of the sensor, such as conductance or cell voltage, on the ambient atmosphere. Electroceramic sensors are often divided into two main types, according to their operation principle, the voltage cell type and the semiconductor sensor type. The zirconia sensor represents the voltage cell type, and this type of sensor is generally used for the measurement of the concentration of oxygen in flue gas. A similar type of sensor is also being developed for the measurement of sulphur dioxide. Semiconductor sensors have been studied and developed for a long time, and they have been successfully tested in combustion gases from power plants, for instance.

An electroceramic solid state sensor has the advantage that it can be placed inside the gas duct. However, it must be protected against the impact of particles by using a porous ceramic filter. The response time of the sensors is generally shorter than the corresponding response time of most gas analysers, because no external sampling line is needed.

7.2.1. Measurement of Oxygen Concentration by Zirconia Cell

In stabilised zirconia (ZrO_2) the oxygen ions can move through anion vacancies. Anion vacancies are formed when lower valence ions (e.g. Y^{3+}) of the stabilising agent are substituted for Zr^{4+} ions. The electronic conductivity of the stabilised zirconia is, moreover, extremely low, so that its ionic transport number t_{ion}, which is the ratio of the ionic conduction to the total conduction, is almost unity.

Figure 7.3 shows the operation of an oxygen sensor formed by a zirconia cell. There the oxygen partial pressures on different sides of a thin zirconia plate are $p_{O_2}^1$ and $p_{O_2}^2$. Both surfaces of the plate are coated with porous electrode material (commonly platinum) which promotes the electrochemical reaction

$$O_2 + 4e^- \rightleftharpoons 2O^{2-} \tag{7.1}$$

where e^- is represents an electron and O^{2-} an oxygen ion [122]. The partial pressure of oxygen determines the chemical potential of the oxygen gas, μ_1 or μ_2 at both electrodes.

The galvanic potential E of the cell depends on chemical potentials on the different surfaces of the zirconia cell as follows:

$$E = -\frac{1}{ZF} \int_{\mu_1}^{\mu_2} t_{ion}\, d\mu \tag{7.2}$$

where Z is the valence of the oxygen ion ($Z = 2$), F is the Faraday constant ($9.65 \times 10^4\,C\,mol^{-1}$), and t_{ion} is the transport number of the oxygen ions.

The chemical potential μ is for its part determined by the partial pressure of oxygen according to the equation [123]

$$\mu = \mu_0 + \tfrac{1}{2}RT \ln p_{O_2} \tag{7.3}$$

where R is the gas constant ($8.309\,J\,mol^{-1}\,K^{-1}$), T is the absolute temperature and μ_0 is a constant. By combining these two equations we obtain

$$E = -\frac{RT}{4F} \int_{p_{O_2}^1}^{p_{O_2}^2} \frac{t_{ion}}{p_{O_2}}\, dp_{O_2} \tag{7.4}$$

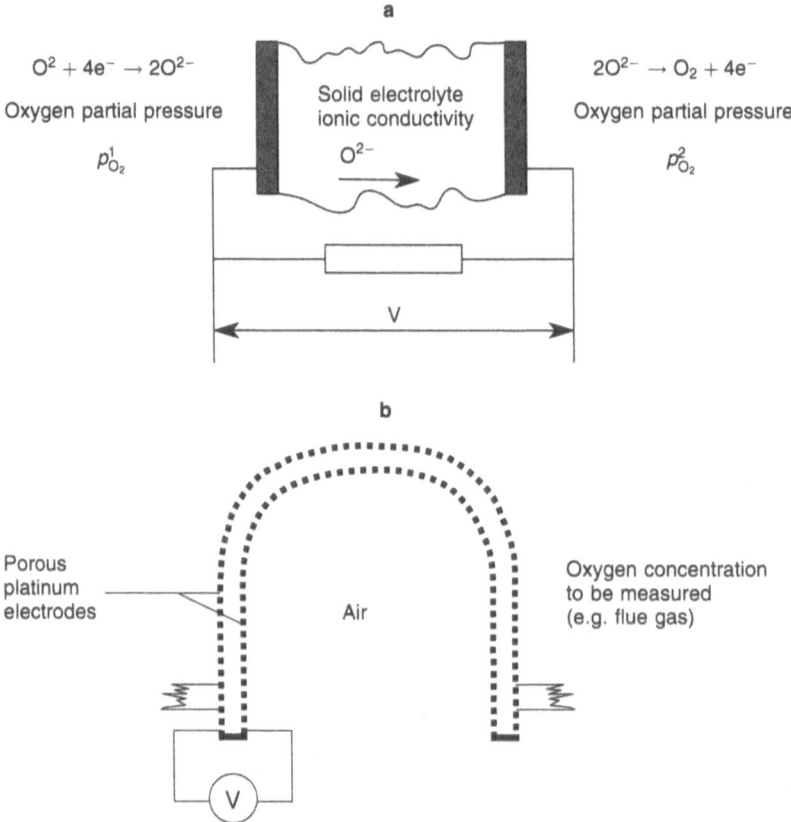

Figure 7.3. **a** Schematic drawing and **b** measurement arrangement of the zirconia sensor.

By taking into account that the transport number t_{ion} of the oxygen ions in stabilised zirconia is, being practical and accurate, unity, we obtain the Nernst equation

$$E = \frac{RT}{4F} \ln\left(\frac{p_{O_2}^1}{p_{O_2}^2}\right) \tag{7.5}$$

The ionic conductivity of zirconia is proportional to the term $\exp(-U/kT)$ where U is the activation energy and k is the Boltzmann constant. In order to operate the cell sensor thus usually requires a temperature of over 500°C.

When measuring oxygen concentrations the reference gas used is normally air, so that $p_{O_2}^2 = 0.2065$ atm. By means of a zirconia cell oxygen concentrations in a flue gas can be measured ranging from a few tenths of one per cent to a few dozen per cent. The dependence of the cell voltage of a certain oxygen sensor on the concentration of oxygen is shown in Fig. 7.4.

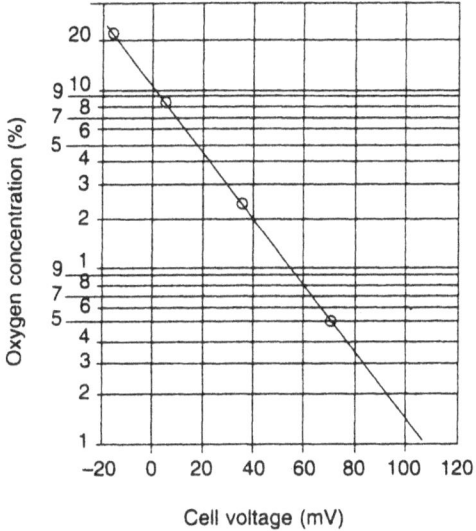

Figure 7.4.   Dependence of the cell voltage of a zirconia sensor on the oxygen concentration [124].

7.2.2. Measurement of Sulphur Oxide Concentrations by Sensor Cells

Solid state electrolytic cells can also be used for the measurement of sulphur oxides. For example, in the case of sulphur dioxide the cell voltage generated is

$$E = \frac{RT}{2F} \ln \left(\frac{p^1_{SO_2} p^1_{O_2}}{p^2_{SO_2} p^2_{O_2}} \right) \tag{7.6}$$

where $p^1_{SO_2}$ and $p^2_{SO_2}$ are the partial pressures of sulphur dioxide on different sides of the cell [122]. When the equilibrium according to the reaction

$$SO_2 + \tfrac{1}{2}O_2 \rightleftharpoons SO_3$$

is realised, the equation

$$E = \frac{RT}{2F} \ln \left(\frac{p^1_{SO_3} \sqrt{p^1_{O_2}}}{p^2_{SO_3} \sqrt{p^2_{O_2}}} \right) \tag{7.7}$$

can be written for the influence of SO_3, where $p^1_{SO_3}$ and $p^2_{SO_3}$ are partial pressures of SO_3 on different sides of the cell. To separate the influence of oxygen from (7.6) and (7.7) a zirconia cell can be used.

The basis for the development of an electrochemical sulphur oxide cell is the selection of a suitable solid electrolyte. Sulphur oxides must react electrochemically and reversibly with the electrolyte. A natural starting point is offered by sulphates, of which the sulphates of alkali metals such as K_2SO_4, Na_2SO_4 and Li_2SO_4 have proved the most appropriate [125–127]. For these the electron conductivity is low, and reversible reactions are possible. M. Gauthier et al. used platinum electrodes with potassium sulphate, and their measuring cell was as follows [126]:

$$(SO_2 + O_2)^1, \ Pt|K_2SO_4|Pt, \ (SO_2 + O_2)^2$$

where the superscripts 1 and 2 refer to each side of the cell. Here the reversible reaction can be represented by the following reaction equation:

$$SO_3(g) + \tfrac{1}{2}O_2 + 2\,e^- \rightleftharpoons SO_4^{2-}$$

The platinum of the electrodes acts as a catalyst ascertaining that SO_3 is an equilibrium concentration in the gas to be measured. This means that the proportions of concentrations between SO_2, SO_3 and O_2 are determinable.

In order to improve the practicability of sulphur oxide sensors, reference electrodes of a solid state have been developed for them, so that the use of a reference gas can be avoided. A solid state electrode with an electrolyte composed of a mixture of lithium sulphate and silver sulphate has, according to the literature, proved reasonably stable and reliable [125].

The measurement range of electrochemical cells extends from about 10 ppm (volume parts per million) to several per cent for sulphur oxides. Their operation temperature is usually around 800°C. The response time of the cell is short.

7.2.3. Ceramic Semiconductor Gas Sensors

The operation of oxide semiconductor gas sensors is based on the change in their conductance, which is caused by a gas reaction that takes place on the surface of the sensor material. The indication of the concentration of the gas to be measured is done by measuring, for instance, the current or voltage, and no reference gas or electrode is needed.

Semiconductor gas sensors have long been used for the detection of leakages of explosive and poisonous gases, for example, in mines, the food industry, and the chemical and metallurgical industries. Research results exist, however, for their use for measuring concentrations of process gases and even for air quality measurements [128–134]. These results are promising and justify optimistic expectations concerning the application of semiconductor sensors to the continuous measurement of emission gas concentrations from combustion processes and gas concentrations in the environmental air. The benefits of this type of sensor are small size, resistance in severe conditions and economy.

Conductance

In the following the conductance properties of semiconductor gas sensors are discussed using as an example tin dioxide (SnO_2), which is the most commonly used basic material of semiconductor gas sensors. Other sensor materials used include zinc oxide (ZnO) and tungsten oxide (WO_3).

Tin dioxide is an *n*-type semiconductor, with an energy gap width of about 3.6 eV, and a semiconductance which is based on oxygen vacancies acting as donors. In semiconductor gas sensors this material is used as small sintered particles, and the microstructure formed by them is shown in Fig. 7.5. As to the current flow in the sensor, the boundaries of the particles form potential energy barriers, which as a whole form a random barrier network.

The gas-sensitive material in gas sensors has an area of a few square millimetres and it is a sintered layer, a few dozens of micrometres in thickness, on a ceramic substrate (Fig. 7.6). In the sensor of Fig. 7.6(a) it is on the outer surface of a thin tube. On the other side of the ceramic substrate there is a heater resistor, as the operating temperature of the sensor is 300–500°C.

The electrical resistance of a semiconductor gas sensor is assumed to be based on the oxygen adsorbed on the surface of the particles. Part of this oxygen is ionised so that electrons originated from the sensor material are trapped in surface states associated with oxygen molecules. This means that O^{2-} ions and possibly also O^- ions are formed. The net charge of the surface so formed generates a potential energy barrier on the surface which resists further

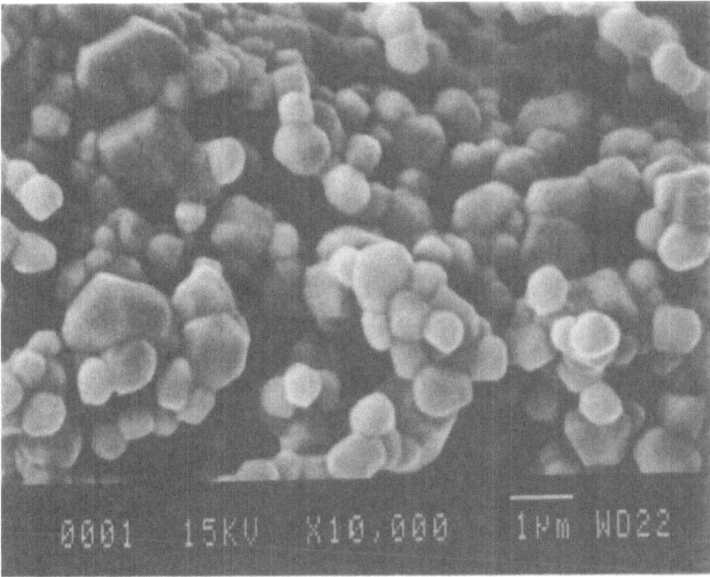

Figure 7.5. Microstructure of gas sensor material. The boundary layers between the particles form a resistance network.

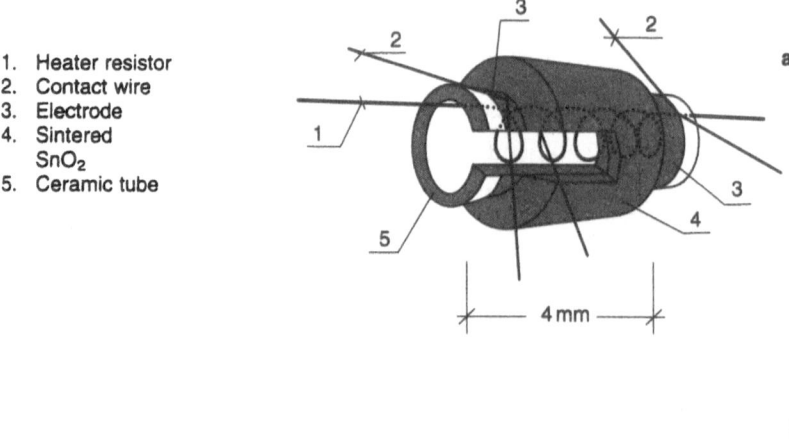

1. Heater resistor
2. Contact wire
3. Electrode
4. Sintered SnO₂
5. Ceramic tube

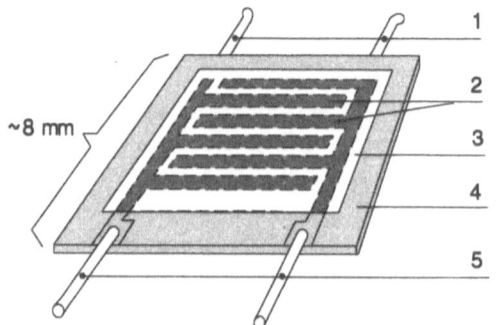

1. Platinum wires for heating resistor
2. Electrodes
3. Gas sensitive layer
4. Al₂O₃ substrate
5. Platinum wires

Figure 7.6. Semiconductor gas sensors. **a** tubular; **b** thick film.

transfer of electrons to surface states. The situation is shown in Fig. 7.7; at the top the potential energy barrier is represented as an upward bend of the lower edge of the conductance band (E_C) and the upper edge of the valence band (E_V) at the boundary surface of the particles. This energy barrier must be crossed by the charge carriers (electrons) in order that the current would flow between the particles.

According to Morrison [135] the conductance G of the sample, formed by pressing of the n-type of semiconductor, at temperature T can be represented by the equation

$$G = G_0 e^{-eV_s/kT} \tag{7.8}$$

where eV_s is the height of the energy barrier (Schottky barrier) between the particles, G_0 is the conductance of material without energy barriers on the particle boundary surfaces, and k is the Boltzmann constant. The energy barrier of the surface is caused by ionised oxygen in which electrons from the semiconductor material have been trapped. Based on this equation Clifford

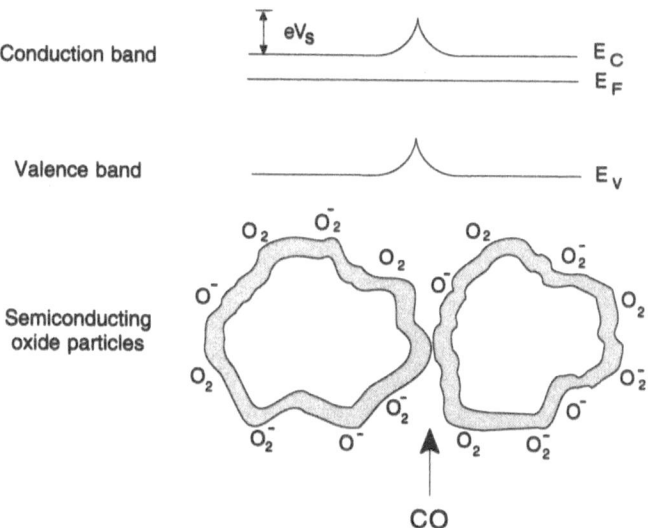

Figure 7.7. Formation of a potential energy barrier on the surface of the particles of the gas sensor material. The shaded surface region refers to a layer depleted from electrons, or the so-called depletion layer. E_F is the Fermi level.

[136] has developed a model of the effect of oxygen on the conductance of the sensor. According to this model, the conductance follows the power-law dependence on the oxygen partial pressure as observed in measurements:

$$G \propto p_{O_2}^{-\beta} \tag{7.9}$$

where p_{O_2} is the partial pressure of oxygen and β is a factor dependent on temperature, normally 0.25–0.5 in magnitude.

Influence of Gases on Conductance. A change in the conductance caused by reducing gases has been observed, similar to oxygen, to follow the power law as a function of the concentration. The power constant is then positive. This can be explained based on catalytic oxidation:

$$n_i G_i + O_{2surf} \longrightarrow G_{prod}$$

In the reaction formula G_i is a reducing gas, n_i is the coefficient of the reaction equation, O_{2surf} oxygen adsorbed on the surface, and G_{prod} the desorbing gas developed as a result of the reaction. When all reactions corresponding to the foregoing reaction equation are taken into account, the conductance can be represented in the power form according to the theory:

$$G \propto p_{O_2}^{-\beta}(1 + \sum_i K_i p_i^{n_i})^{\beta} \tag{7.10}$$

where p_is are the partial pressures of the various gas components, and K_is so-called sensitivity coefficients. The sensitivity and selectivity of a semiconductor gas sensor with regard to a certain gas component can be improved, if the sensitivity coefficient of the sensor to this component can be increased compared with other gases. Means to achieve it include the modification of the composition of the sensor material by additives, or the optimisation of the operation temperature of the sensor.

For a commercial tin dioxide sensor, sensitivity coefficient values have been measured [136]: $K_{CH_4} \approx 5 \times 10^{-3}\,ppm^{-1}$, $K_{H_2O} \approx 5 \times 10^{-3}\,ppm^{-1}$, $K_{H_2} \approx 0.1$ ppm^{-1} and $K_{CO} \approx 1 \times 10^{-3}\,ppm^{-1}$. This sensor is very sensitive to hydrogen.

Use of Electroceramic Gas Sensors

The zirconia cell sensor is generally used for the measurement of oxygen concentrations, for example from flue gas. Semiconductor gas sensors are used today for the detection of leakages of inflammable, explosive, or toxic gases in mining establishments, food industry, as well as chemical and metallurgical industry.

Semiconductor gas sensors have been tested for years for the measurement of concentrations of flue gases. Research into the applicability of this type of sensor has mainly been concerned with measurement of the carbon monoxide concentration in the flue gas. According to tests, sensors follow the concentration of carbon monoxide in the flue gas, but their stability needs to be improved.

Figure 7.8 shows a response of a semiconductor (thick film) sensor to the concentration of carbon monoxide in a combustion gas. The sensor follows the concentration well. Figure 7.9 shows the improvement in the sensor performance achieved by the introduction of a catalytic additive. As can be noticed, the catalytic agent produces a faster, more detailed response.

Semiconductor sensors have in some tests proved suitable for measuring the concentration of hydrogen sulphide. Semiconductor gas sensors have also been tested to measure nitrogen oxides, primarily nitrogen monoxide, from the flue gas. For air quality measurements, both lead phtalocyanine (for nitrogen dioxide) and tin dioxide (for carbon monoxide) have been tested.

Selectivity

As was noticed (Fig. 7.9), the performance of semiconductor gas sensors can be varied by adding catalytic agents to the sensor material. These can control sensitivity, selectivity and stability characteristics of sensors. Their influence is dependent on the distribution of the additive in the material or at the surface of the particles, which form the microstructure of the sensor. The distribution can be controlled by heat treatments. Sensitivity and selectivity of sensors can also be adjusted by incorporating other additive materials, such as different metal oxides.

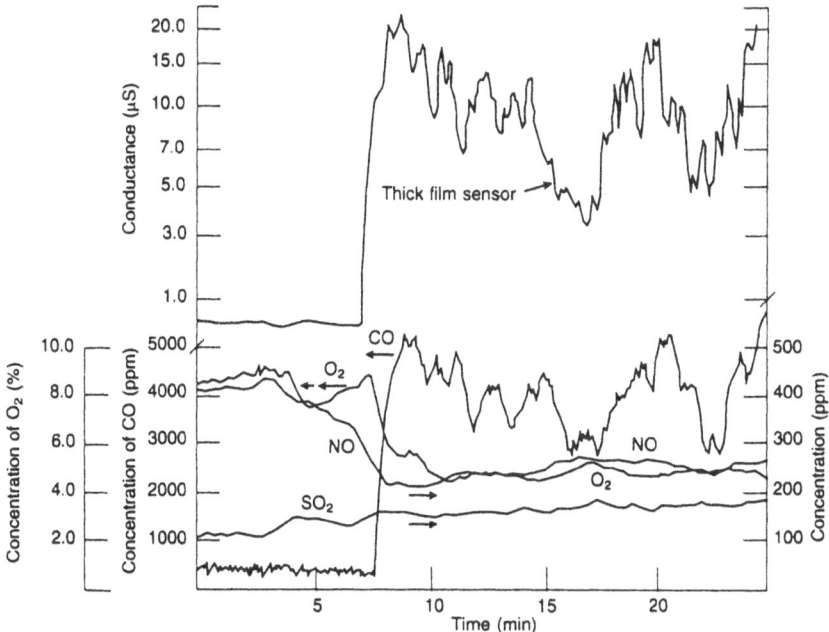

Figure 7.8. Conductance response of a thick-film gas sensor (upper curve), and the concentrations of carbon monoxide, oxygen, nitrogen monoxide and sulphur dioxide from the combustion gas of sod peat. The experiment was performed on an experimental heating power station for research and professional training. The nominal power of this unit was 1.1 MW [131].

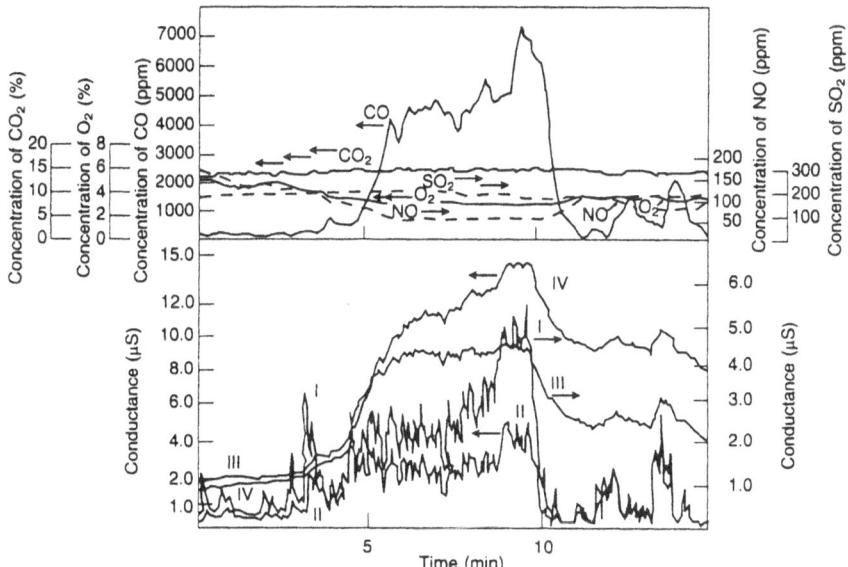

Figure 7.9. Conductance response of three types of thick-film sensors from the combustion gas of milled peat, as measured on a city power station. Sensors I and II contained palladium as a catalytic additive. Sensor III contained platinum, whereas sensor IV contained no catalytic additive. Concentrations of CO, CO_2, O_2, NO and SO_2, as measured using reference instruments, are also presented [129].

The control of the sensor operation temperature can also be used to control the performance characteristics of semiconductor sensors, such as the selectivity and the response time. In some cases, the cycling of the sensor temperature between two values has resulted in an increase in the selectivity and in a decrease in the response time. Such an experiment is shown in Fig. 7.10. The figure shows the time dependence of the conductance of a thick-film sensor when temperature was cycled continuously between 400 and 150°C in synthetic air containing different concentrations of H_2S [137]. The cycle length was $60 + 60$ s. At zero concentration of H_2S, the conductance decreases about a thousand-fold after cooling from 400 to 150°C. As shown in Fig. 7.10, small concentrations of H_2S change the form of the conductance response drastically in the case of continuous temperature cycling. Test results illustrate the high sensitivity of the thick-film sensor tested to H_2S after a cooling process. When 1 ppm of H_2S was present in the ambient gas, only about a fifty-fold decrease in conductance occurred after the cooling. A marked increase in conductance appeared at 150°C after cooling from 400°C at concentrations of 10 ppm and higher. The characteristic features of the conductance response curves shown in Fig. 7.10 can be related to corresponding H_2S concentrations. Stability characteristics of the sensor type were also investigated. Temperature cycling measurements were continued over a period of four months, but no changes were observed in the form of conductance responses.

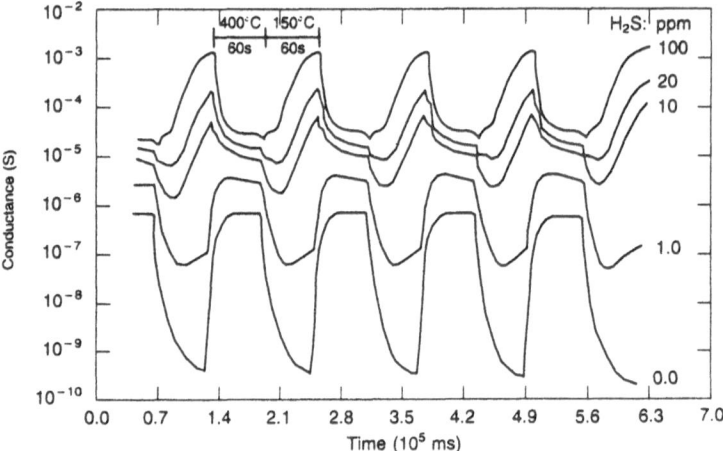

Figure 7.10. Time dependence of the conductance of a thick-film sensor when temperature is continuously cycled between 400 and 150°C with a cycle length of $60 + 60$ s in synthetic air containing 0, 1, 10, 20 and 100 ppm of H_2S [137].

7.2.4. SAW Gas Sensors

Surface acoustic wave (SAW) sensors rely on the interaction of a gas with a specially chosen and prepared layer [138]. A change of the gas concentration results in a change of the mass of the layer which affects the acoustic field of the SAW, and in the change of the electrical conductivity of the layer which affects the electric field of the SAW associated with the acoustic field. These changes influence the signal flow of the device. The interaction with the gas usually has to be selective and reversible, resulting in a sensitive sensor with a fast response time. In addition, the chemical interface has to be stable at elevated temperatures (above 100°C) which prevent the condensation of water and reduce response time.

In SAW sensors, the concentration measurement of a specific gas in a mixture of gases can be performed using a chemical interface [138]. This is a chemical or biochemical compound which interacts selectively with the measurand. During this interaction, the physical properties of the chemical interface are changed and these can modulate a signal flow, e.g. a flow of charge carriers. The change in the physical properties modulates the SAW phase velocity. Organic semiconductors, such as metal-free phthalocyanines, can be used for the chemical interface.

The sensor system can be based upon a dual delay-line oscillator, where the difference between the two oscillator frequencies is a measure of the gas concentration. A sensitivity of about $100\,\text{Hz}\,\text{ppm}^{-1}$ has been obtained for NO_2 giving a threshold sensitivity of about 0.5 ppm [138].

7.3. Generalised Theory of Gas Sensors

Solid state gas sensors based on the interaction between an electrical system (the sensing element) and a chemical system (the measurement gas) are used extensively to measure a variety of gases including O_2, CO, H_2 and hydrocarbons [139, 140]. The understanding of the operation of these gas sensors has improved and specialised models describing sensor operation have been developed. More comprehensive models still need to be developed, such as a first-principles steady state model for the TiO_2 gas sensor, capable of explaining many of the experimental features of that sensor [141]. The generalised model relates the electrical response (e.g. resistivity, electrochemical potential) of a metal oxide to the concentration of a species in the measurement gas [139]. A general idea about the application of the model to the TiO_2 resistive type sensor and the ZrO_2 voltage cell type sensor is given.

7.3.1. General Model

The metal oxide sensing element is assumed to be homogeneous and non-porous and contain a certain concentration c_v of mobile oxygen vacancies whose charge $(+2e)$ is compensated by mobile electrons of concentration n and by immobile dopants of concentration c_d. The vacancies can diffuse to the surface where they combine with electrons and adsorbed oxygen atoms to produce perfect lattice sites [139]. The reverse reaction can also happen.

$$v + \Sigma \underset{k_1}{\overset{k_0}{\rightleftharpoons}} v_s$$

$$v_s + 2e + O_{ads} \underset{k_F}{\overset{k_A}{\rightleftharpoons}} 2\Sigma$$

where v, v_s and Σ denote a vacancy in the interior, a surface vacancy, and a surface lattice site, respectively. The electrons and adsorbed oxygen atoms may originate from the bulk material (TiO_2) or from an electrode (ZrO_2). The adsorbed reducing species (e.g. CO, H_2, CH_4) in the ambient gas react with adsorbed oxygen generating vacancies. The gas–solid interface thus acts as a source or sink for oxygen vacancies (and electrons) resulting in a change in the charge concentration or the charge distribution inside the metal oxide.

One has to formulate and solve two parts of the sensing process [139]. The first is the electrical system, which may be described by the equations for the electric current for each charged species, and for the electrostatic potential.

The second part of the sensing system is the chemical system which involves motion of gas molecules to and from the solid surface, and adsorption, desorption and reaction processes on the surface. As an illustration, the measurement gas is chosen to consist of only O_2, CO and N_2 with the following simple reaction scheme:

$$O_{2g} + \Sigma \underset{k_1^d}{\overset{k_1^a}{\rightleftharpoons}} O_{2ads} \qquad\qquad O_{2ads} + \Sigma \underset{k_R}{\overset{k_D}{\rightleftharpoons}} 2O_{ads}$$

$$CO_g + \Sigma \underset{k_2^d}{\overset{k_2^a}{\rightleftharpoons}} CO_{ads} \qquad\qquad CO_{ads} + O_{ads} \underset{k_b}{\overset{k_f}{\rightleftharpoons}} CO_{2ads} + \Sigma$$

For steady state, the fractional occupancies of surface sites by different species and the net transfer rate of these species from the bulk of the gas phase (by diffusion or by flow) can be interrelated. The solution of the equations so obtained provides the value of the parameter θ_0, which is dependent on the coefficients k_0, k_1, k_A, k_F, and on the concentrations of conducting electrons and mobile oxygen vacancies. This parameter (θ_0) determines the electrical response of the sensor.

7.3.2. Resistive-Type Sensors

The application of the model to the resistive type metal oxide gas sensor (TiO_2) can explain the operation of this sensor type as an oxygen sensor and as a carbon monoxide sensor. Under local thermodynamic equilibrium conditions, the resistivity of TiO_2 in an O_2–CO–N_2 gas mixture is determined by the equilibrium oxygen partial pressure and shows a large stepwise change at the stoichiometric composition ($R_c = p_{CO}/p_{O_2} = 2$) [139]. In the absence of thermodynamic equilibrium, however, the step can occur at any value of $R = p_{CO}/p_{O_2}$ depending on the values of the adsorption and desorption rate constants for O_2 and CO. Also in this case, and in the limit of very small values of R, the resistivity can show a *power law dependence* on p_{CO}. Under these conditions, the metal oxide functions as a CO sensor. The effect of adding catalytic particles to a metal oxide as well as the effect of porosity can be included in the generalised model [141].

7.3.3. Electrochemical Zirconia Sensor

An electrochemical zirconia (ZrO_2) sensor (voltage type sensor) consists of an yttrium stabilised zirconia cell with porous platinum electrodes on the two surfaces of the cell. The presence of yttrium results in a high concentration of mobile oxygen vacancies c_v which is much higher than the concentration of electrons [139].

Similar to the case of the resistive-type sensor, the emf (open circuit voltage) of the zirconia cell at thermodynamic equilibrium is defined by the oxygen partial pressure and exhibits a stepwise change at the stoichiometric CO/O_2 ratio. In the absence of thermodynamic equilibrium, the step occurs at different ratios depending on the motion of gas molecules to and away from the solid surface, on adsorption and desorption, and on reaction processes at the lattice sites on the surface.

7.4. Integrated Arrays of Multiple Sensors

The analysis of emission gases and air impurity gases requires the concentration of many gas components to be measured. Sensor technologies are being developed in order to improve the selectivities of individual sensors. There is also another approach, which combines several sensors together and uses the chemometric analysis of the responses from individual sensors. Instead of high selectivity, this analysis method utilises the differences between the responses of different sensors.

Solid state gas sensors have been used in the development of sensor arrays. Thin film sensors are an obvious choice for the construction of sensor arrays, as these can be integrated using standard planar circuit technologies. In a study, eight thin film metal oxide sensors were integrated together for the analysis of mixtures of gases and vapours [142]. The sensing elements were constituted using single layers and bilayers of reactively sputtered thin films of SnO_2, ZnO, TiO_2, and WO_3 with and without Pd catalyst. These had different cross-sensitivities to species of interest. Sensor temperature could be controlled by on-chip boron-diffused heating elements. The sensor response spectra were processed by an artificial neural network utilising response patterns obtained from the sensor array. The sensitivity of a sensor array is dependent on the stand-alone characteristics of the single sensors and on their mutual interaction [143].

7.5. Measurement Methods Based on Catalytic Combustion

To measure the concentration of combustible gases a method based on the catalytic combustion can be used. It usually consists of two temperature sensors, the operation of which is based on the change of the electrical resistance. One of them is covered with a catalytic agent, e.g. platinum, and the other with an inert material. The unburnt gas components remaining in the flue gas are burned due to the catalytic agent, raising the temperature of the sensor covered with the catalyst. The temperature difference between the measuring and reference electrodes is then dependent on the concentration of combustible gases and can be used as a measurement signal.

7.6. Development Needs and Possibilities in Emission Measurement Technologies

The technologies used for the measurement of concentrations of emission gases were developed mainly in the last few decades. Consequently, they are the object of continuous research and development. For instance, for the measurement of solids emissions, a continuously operating measuring device would be needed to complement the method based on periodical sampling generally used today.

For the measurement of gaseous emissions, optical measuring principles are applied and developed at the moment, as for example the techniques based on the interpretation of spectra obtained by means of FTIR technology. This technology also requires effective data processing and mathematical treatment. In addition to the analysis of the individual spectra, computer programs are being developed to perform multicomponent analyses. By means of FTIR technology, it is now possible to determine the concentrations of several emission gas components almost continuously.

The development of different sensors to measure gas concentrations directly from the gas channel or through a sampling line is also in progress in several dozens of research institutes in various parts of the world, with an emphasis on continuous measuring methods. These kinds of methods are necessary so that the control of processes can be made more efficient and that the quality and quantity of the total emissions from processes can be elucidated.

Chapter 8
Analysis Methods

In the following, the basis of some important methods for the analysis of gaseous emissions and particulate emissions are discussed. These methods include gas chromatography, mass spectrometry, atomic absorption spectrometry and X-ray spectrometry.

8.1. Gas Chromatography

Gas chromatography is a very versatile gas analysis method. The components of a gas sample can be separated by means of the column of a gas chromatograph [72]. This selectivity is the greatest advantage of gas chromatography. However chromatographic analysis takes at least several minutes, and it cannot therefore be considered as a continuous method.

Gas chromatography is used to separate gases from each other. The central part of a gas chromatograph is a long tube or a column which, in the case of a capillary column, is usually coil-shaped, because its length must be several metres or dozens of metres. A packed column, a few millimetres in diameter, is filled with large-area solid support, the surface of which is covered with a layer formed by a liquid which will not vaporise easily, such as esters or silicone oils with high boiling temperature. This is the stationary phase. The filling materials of the columns most frequently used are molecular sieves, porous polymers, silica gel and alumina. The separating effect of the filler is based on the adsorption or the sieve effect. The column is thin in order not to require large amounts of materials. The length of a packed column is from a few dozens of centimetres to a few metres. Today, the most commonly used columns are capillary columns, which are necessary in high resolution gas chromatographs. A capillary column is a tube drawn of quartz, several dozens of metres in length and with an inner diameter of about 0.2 mm [144]. On the inner surface of the column, there is a thin layer that differentially adsorbs the gases to be studied.

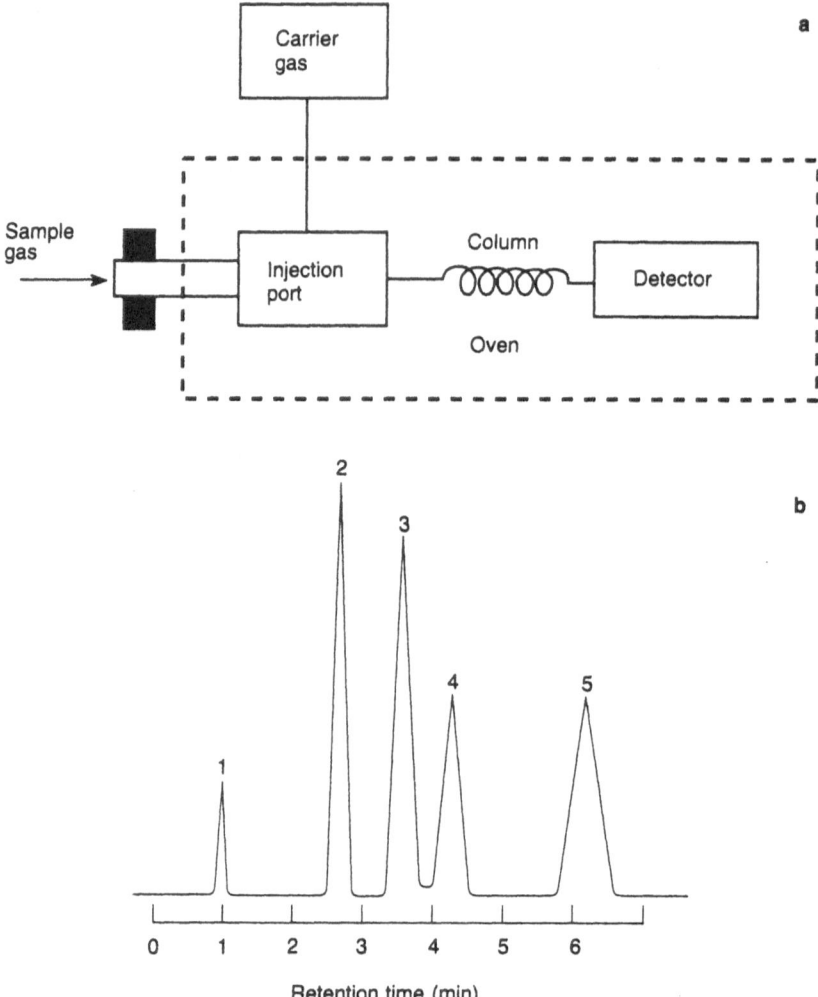

Figure 8.1. **a** Schematic representation of a gas chromatograph. (After H. A. Strobel, *Chemical Instrumentation: a Systematic Approach*, 2nd edn, © 1973 Addison-Wesley Publishing Company.) **b** A chromatogram representing the separation of different gas compounds as a function of time.

Pure carrier gas, which is normally nitrogen or helium, is driven through the column. A small amount of the material to be studied is injected in the flow of the carrier gas, and this material starts to travel with the carrier gas.

The materials in the column retard the various components of the sample gas that travel with the carrier gas, in different ways. This retardation is affected by the affinity of the components of the sample gas with the stationary phase and to the mobile phase. Differences in the affinity, and the separation between the various phases due to these differences, are mainly based on adsorption forces and ion exchange phenomena at the phase boundaries. The balance between the various phases based on adsorption and separation phenomena depends on the properties of these phases and the components of the sample gas [145]. Figure 8.1(a) [72] shows the principle of a gas chromatograph.

Many materials to be analysed using a gas chromatograph do not evaporate sufficiently at the ambient temperature of a laboratory, and would not migrate through the column at this temperature. Therefore the gas chromatographic analysis is often performed at an elevated temperature. The temperatures vary ordinarily between 100°C and 300°C, and the increase of the temperature can be programmed as a function of time. Some stationary phases tolerate a temperature of even 450°C without decomposing.

Multiple column systems with automatic valves between them must often be used in order that the components of the sample gas can be separated effectively enough.

8.1.1. Retention Time

The basis of gas chromatography is the separation of the sample gas components from each other. The separation is controlled by the distribution balance of the gas component between the mobile phase and the stationary phase. A material more strongly adsorbed at the stationary phase remains in the column for a longer time than a material more weakly adsorbed, i.e. the latter runs more rapidly through the column. The time it takes for the gas component to travel through the column, that is, the time between injection and detection, is called the retention time Fig. 8.1(b). In the determination of the retention time of gases, calibration gases are utilised. The operation of a gas chromatograph presupposes therefore that the gas component examined is detected when it exits the end of the column. Various kinds of detectors have been developed for this purpose.

8.1.2. Chromatograph Detectors

The function of the detector is to indicate the gas component to be measured among the carrier gas utilising some chemical or physical phenomenon. The first detectors were based on heat conduction, which could be used to detect the gas studied from the carrier gas stream. To detect organic compounds, the flame ionisation detector (FID) based on hydrogen flame was developed. It could detect gas concentrations of a few ppm. Numerous other detectors have since been developed.

Detector Based on Thermal Conductivity (TC Cell)

The heat transfer from a resistor wire to the gas depends on the thermal conductivity of the ambient gas. The greater the thermal conductivity, the more the resistor wire is cooled, and its resistance decreases. Detectors exploiting this phenomenon are represented in Fig. 8.2, and thermal conductivities of various gases compared with that of air are shown in Table 8.1. By using hydrogen or helium as the carrier gas, a great difference in heat conduction is achieved compared with other gases. A detector based on thermal conductivity is not selective, but since the chromatograph separates the gas components, it can be used provided that the retention times of the column have been calibrated.

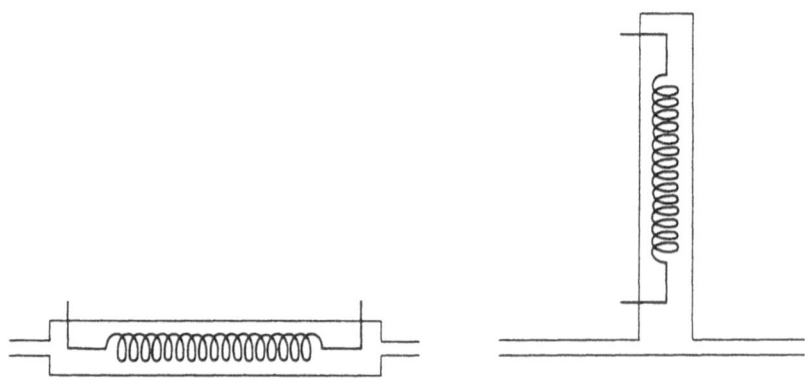

Figure 8.2. Detectors of a gas chromatograph based on thermal conductivity [90].

Table 8.1. Thermal conductivities of gases compared with that of air [90]. The thermal conductivity of air is $296\,\mu\mathrm{W\,cm^{-1}\,K}$

Gas	Chemical formula	0°C	100°C
Acetylene	C_2H_2	0.78	0.90
Air	–	1.00	1.00
Ammonia	NH_3	0.90	1.04
Argon	A	0.68	0.70
n-butane	C_4H_{10}	0.55	0.74
Carbon dioxide	CO_2	0.61	0.70
Carbon monoxide	CO	0.96	0.96
Chlorine	Cl_2	0.32	–
Ethane	C_2H_6	0.75	0.97
Helium	He	5.97	5.55
Hydrogen	H_2	7.15	6.90
Krypton	Kr	0.36	–
Methane	CH_4	1.25	1.43
Methyl bromide	CH_3Br	0.26	–
Neon	Ne	1.90	1.8
Nitrogen	N_2	0.995	0.996
Oxygen	O_2	1.013	1.014
Sulphur dioxide	SO_2	0.35	–
Water vapour	H_2O	–	0.78
Xenon	Xe	0.21	–

Flame Ionisation Detector

This detector is similar to that used for the measurement of hydrocarbons. It uses a hydrogen flame and it detects all organic materials which have a bond of carbon and hydrogen. The detector is very sensitive.

Flame Photometric Detector

In the presence of sulphur-containing compounds the hydrogen flame emits light, the wavelength of which is 394 nm. To detect this light, a narrow-band filter and a photomultiplier tube are used. The structure of this flame chamber is similar to that of the flame ionisation detector (Fig. 8.3), but operation conditions can be varied. The flame photometric detector is often used for the detection of small sulphur concentrations, as in the analysis of mercaptans from emission gases.

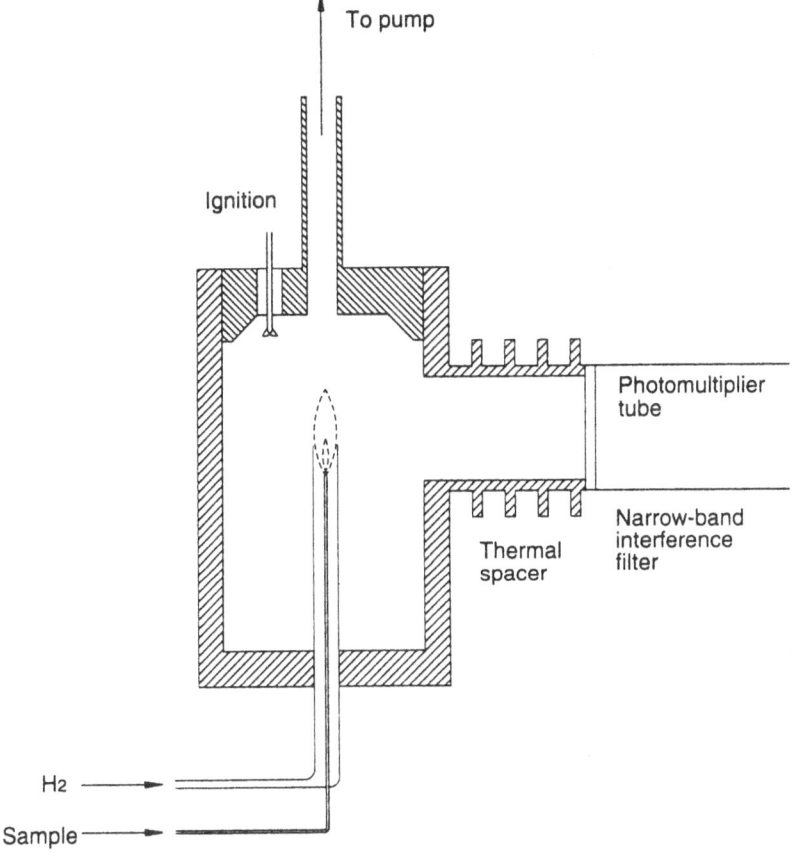

Figure 8.3.   Flame photometric detector [90].

Electron Capture Detector

In an electron capture detector, the source of ionising radiation (for example, ^{63}Ni in a nickel foil) produces electrons which are collected by using just a sufficiently high potential. Molecules entering the measuring chamber with the carrying gas which have a high electron affinity, capture electrons. Therefore a decrease is observed in the current measured, which indicates the arrival of the compound in the chamber. By using the electron capture technique, oxygen, sulphur and halogens, for instance, can be detected. It is also commonly used for the detection of organic compounds of emission gases.

Modern emission analysis also uses mass spectrometers and FTIR analysers as effective detectors in gas chromatography. These form a combination of two analytical devices. There is also equipment which has a combined gas chromatograph, FTIR analyser and mass spectrometer as one whole. These are versatile analytical devices which are used, for example, for the analysis of complex harmful substances. These can contain isomers of harmful substances such as dichlorobenzenes, dinitrotoluenes, PCB compounds and dioxins [146]. For example, the mass spectra of the isomers 1,2-dichlorobenzene and 1,4-dichlorobenzene are practically identical. Their infrared (FTIR) spectra differ, however, considerably from each other and the isomers can be detected unambiguously on the basis of these spectra.

8.2. Mass Spectrometry

Mass spectrometry is an analysis technique in which the material is converted into a gaseous form, ionised and analysed on the basis of the ratio of the mass and charge [72]. The technique can be used for both qualitative and quantitative analysis.

Gas, or material which has been brought into a gaseous state by heating, is ionised in the ion source. The ions produced are arranged in the mass analyser with regard to mass and charge, that is to say, in accordance with the m/e ratio. The ions arranged in the mass analyser are detected in the detector of the mass spectrometer.

Mass spectrometry is a destructive method of analysis, since the sample has to be ionised. A very small amount of material is, however, sufficient for a sample, even an amount of 100 ng. In some cases, mass spectrometry is not a practicable method, because the sample material decomposes with heating.

Today, mass spectrometry is a very important method for the analysis of emissions. Combined with gas chromatography, it is used for the determination of organic emissions. A mass spectrometer then operates as the detector of a gas chromatograph. A high-resolution gas chromatogaph plus mass spectrometer is also suited to the analysis of dioxins and furans.

The resolution of a mass spectrometer indicates how well it can distinguish between peaks near each other in the spectrum. The resolution is defined as the ratio $m/\Delta m$, where m is the mass number corresponding to the peak and Δm is the difference in mass units compared with the mass number of the nearby peak. For example, a reasonable resolution 3000 means at the m/e ratio 300 that ions 300 and 300.1 expressed in m/e ratios can be distinguished from each other. The mass numbers of materials to be analysed with a mass spectrometer can typically vary from about ten to hundreds and even thousands.

8.2.1. Production of Ions in Mass Spectrometer

The ions of the sample gas are usually produced by using *electron ionisation*. For example, the mass spectra of organic compounds are produced by ionising the sample gas by means of electrons which are emitted from a tungsten or rhenium filament and which are normally accelerated by a potential difference of 70–80 V. The ionisation energy can be controlled, for instance, between 10 and 250 eV. The pressure of the ionisation chamber is 10^{-6}–10^{-7} torr. In the ionisation, primarily positive ions are developed.

When using electron ionisation, electrons are directed to a small area where there is sample gas. If the sample molecules are denoted with the symbol M and the electron with the symbol e^-, the ionisation process can be represented using the reaction scheme [147]

$$M + e^- \longrightarrow M^+ + 2e^-$$

The energy transferred from the ionising electron to M is sufficient (1) to remove an electron from the molecule, in which case a molecule ion is formed, and (2) to induce fragmentation of some molecules, causing neutral particles and fragment ions to be formed. Electron ionisation is a very common method, but its drawback is the excessive fragmentation of some molecules so that the spectrum only contains peaks of fragment ions, since all molecule ions are split before arriving at the detector.

Chemical ionisation offers a milder method for the generation of ions. In this method, reaction gas is brought to the ionisation region (the ion source). Even in this case, there is underpressure in the ion source, but the pressure is, however, considerably higher than in the electron ionisation. The electrons are again directed through the ion source, but this time they ionise the reaction gas. The most common reaction gases are ammonia, methane and isobutane. In the following, the example used is ammonia:

$$NH_3 + e^- \longrightarrow NH_3{}^+ + 2e^-$$

Molecule ions of ammonia formed originally in the higher pressure of the chemical ionisation source collide with neutral ammonia molecules, and hydrogen transfer typically occurs:

$$NH_3{}^+ + NH_3 \longrightarrow NH_4{}^+ + NH_2$$

When colliding with gas molecules to be analysed, NH_4^+ ions of the reaction gas produce ionisation either by proton transfer

$$M + NH_4^+ \longrightarrow [M + H]^+ + NH_3$$

or by addition

$$M + NH_4^+ \longrightarrow [M + NH_4]^+$$

Chemical ionisation mostly produces protonised molecules $[M + H]^+$, the mass number of which is higher by one than that of the original molecule.

8.2.2. Mass Analysers

The analysis of molecule ions in a mass spectrometer is performed using a mass analyser. According to the two most frequently used methods of analysis, the mass spectrometers are divided into two groups, namely magnetic sector instruments and quadrupole instruments.

Magnetic analysis is based on the fact that a charged particle with a definite velocity entering a magnetic field assumes a circular orbit. The radius of this orbit is determined so that the centrifugal force exactly compensates the force caused by the magnetic field. The radius of the orbit is then determined by the mass and charge of the particle. Particles which have a different mass or charge and which have been accelerated in a uniform electric field, thus have an orbital radius unequal in size in the magnetic field. They can be separated after the magnetic field using a detector. In practice, the separation is done so that the magnetic field intensity is changed or scanned so that all m/e values are detected. The principle of magnetic separation is represented in Fig. 8.4 [72]. The acceleration voltage of the ions is of the order of a few thousand volts.

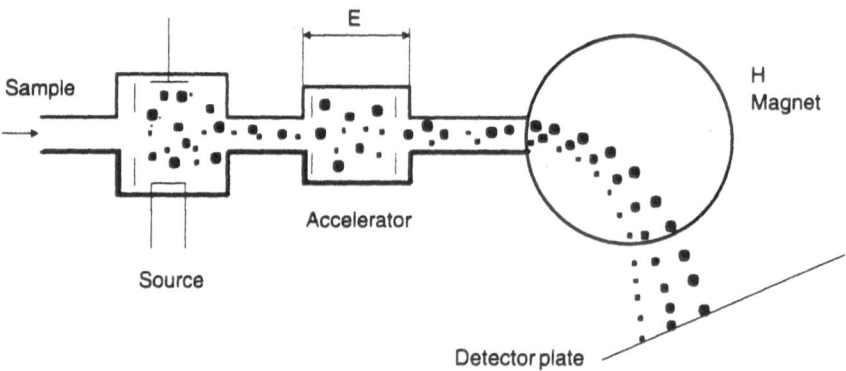

Figure 8.4. Magnetic separation of ions having different m/e values. (After H.A. Strobel, *Chemical Instrumentation: a Systematic Approach*, 2nd edn, © 1973 Addison-Wesley Publishing Company.) The detector plate is only for historical reasons to visualise the principle. Presently, electric detectors are used.

In order to improve the resolution a small slit is used in front of the detector which admits only particles having a radius of a definite orbit. Before the magnetic separation, an electrostatic analyser is commonly used which admits only particles travelling in exactly the right direction. All in all, this means that magnetic analysis is performed from an ion beam which is definite with regard to energy. In other words, the ions entering the slit of the detector have been focused twice, that is, once with regard to direction and once with regard to velocity. Thus a high resolution is achieved.

Another common type of mass analyser is a quadrupole mass filter. It is formed by four parallel bars, which sort out the ions as they travel longitudinally between the bars. One pair of opposite bars is set at voltage V_{dc} and the other pair at voltage $-V_{dc}$ (Fig. 8.5). In addition, a radio-frequency voltage is connected to the first pair of the opposite bars, and a voltage of opposite phase to the other pair. When the ratio of the d.c. voltages and the alternating voltages is suitable and when the bars are sufficiently long, most of the ions entering the quadrupole area oscillate and are thrown against the walls. As the alternating voltage and its frequency have certain values, there exist ions with one m/e value which can reach the analyser and be detected. The whole spectrum can be produced by changing the frequency while the alternating and d.c. voltages remain constant, or by changing the alternating and d.c. voltage simultaneously so that the ratio of these voltages remains the same, keeping the frequency constant. The operation principle of the quadrupole mass analyser is shown in Fig. 8.5 [72]. The length of the quadrupole bars can be of the order of 20 cm and their active surface is hyperbolic in shape.

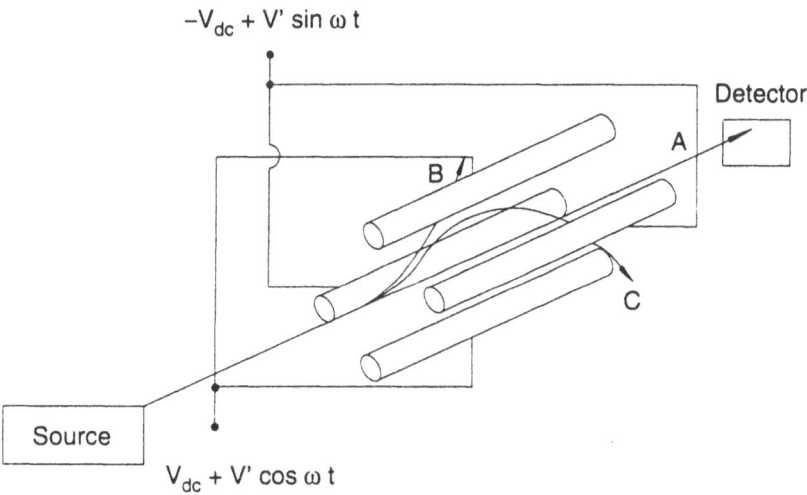

Figure 8.5. Schematic drawing of a quadrupole mass filter. Ion A penetrates the analyser and becomes detected. The m/e ratio of ions B and C is unsuitable for the electrical values of the quadrupole, and these ions are thrown against the walls. (After H. A. Strobel, *Chemical Instrumentation: a Systematic Approach*, 2nd edn, © 1973 Addison-Wesley Publishing Company.)

8.3. Gas Chromatograph–Mass Spectrometer in Combination

A gas chromatograph alone is not capable of reliably identifying the gases to be analysed. It does, in fact, only measure the time it takes for each gas component to travel through the chromatograph column. Several gases can have the same retention time. On the other hand, mass spectrometry is a very efficient technique for identifying components, but the mass spectra obtained from gas mixtures are normally too complex to be used [147]. By combining these two techniques an effective method of analysis is obtained, which exploits both the capability of a gas chromatograph to separate gases from each other and that of a mass spectrometer to identify the compounds separated.

These two techniques are in most respects very compatible. Both use gaseous samples so that a gas, or a sample converted into gas form, to be separated by a gas chromatograph, can directly be conducted to a mass spectrometer. For both techniques, a very small quantity of sample is sufficient (pico- or nanograms). The methods complement each other well, as shown in Table 8.2. One exception can, however, be seen from the table. The outlet of the gas chromatograph is at ambient pressure whereas the mass spectrometer operates well at a very low pressure. This conflict has, nevertheless, been solved technically.

Table 8.2. Gas chromatography and mass spectrometry are complementary to each other as analysis techniques [147]

Property	Gas chromatography	Mass spectrometry
Capability of treating mixtures	Yes	No
Capability of identifying	Not unambiguous	Yes
Phase to be processed	Gas	Gas
Picogram range	Yes	Yes
Operation pressure	Air pressure	High vacuum

The direct combination of a gas chromatograph and a mass spectrometer means that the pressure of the ion source of a mass spectrometer is higher than optimal. The outlet of the gas chromatograph, on the other hand, is no longer at air pressure. This affects the flow of the gas so that the resolution of the chromatograph deteriorates a little. These slight losses in the efficiency of both the gas chromatograph and the mass spectrometer are not significant from the point of view of the analysis.

Combinations of a gas chromatograph and a mass spectrometer are automated and computerised. Of the emission gases, chlorinated dioxins and furans are analysed using a combination of a gas chromatograph and a mass spectrometer.

When analysing polychlorinated dibenzodioxins (PCDD) and polychlorinated dibenzofurans (PCDF), selective ion monitoring (SIM) is often used to separate various isomers from each other. There can, as a matter of fact, be considerable differences in the toxicity of various isomers. The SIM techniques can also be used if the concentrations are very small. A mass spectrometer is tuned in this technique to detect only a definite ion or a fragment ion.

8.4. Atomic Absorption Spectrometry

Atomic spectroscopy refers to spectroscopic methods of analysis in which the sample is brought into an atomic state and the absorption or emission of light induced by the atoms is measured [148]. In atomic absorption spectroscopy, the light is directed through the atomic vapour and the intensity of the light absorbed is measured. The method is generally used for quantitative analysis in laboratories performing inorganic determinations. It is very suitable for the determination of concentrations of numerous substances, and is capable of determining very small concentrations. The operating principle of an atomic absorption spectrometer is shown in Fig. 8.6.

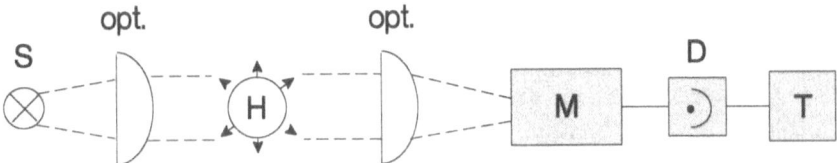

Figure 8.6. Operation principle of an atomic absorption spectrometer [148]. S, light source; opt, optical components; D, light detector; H, vaporiser for the atomisation of the sample; M, monochromator; T, output display.

As shown above, the absorption and emission of light is related to a change in the atomic energy from one level to another, or the energy transition. In the absorption the atom is excited, and in the emission the excited state is discharged. A spectral line where both absorption and emission take place at the same wavelength is called a resonance line. In atomic absorption spectrometry, only resonance wavelengths are used. The width of a typical absorption peak is less than 0.01 nm.

8.4.1. Light Source

The light source used in an atomic absorption spectrometer is generally a line source, or in other words, the light peak of a resonance line of the material. This is practically monochromatic light. The intensity of the peak is also high compared with, for instance, a 0.01 nm band separated from a continuous light spectrum by means of a monochromator and corresponding to the resonance wavelength.

Usually, a so-called hollow cathode lamp is used as the light source. The cathode of the lamp, which has the shape of a hollow cylinder, contains the chemical element to be determined and thus emits the characteristic spectrum of the element. This means that a lamp is normally needed for each element to be determined. The hollow cathode lamp contains argon or neon at the pressure of 1–7 mbar, and naturally another electrode or an anode which is made of tantalum, zirconium, tungsten or nickel. When the voltage between the electrodes is 100–400 V, positive noble gas ions formed in the discharge, sputter or release the material intended from the inner surface of the cathode. This is ionised by collisions with positive ions or secondary electrons originating from the cathode. The material detached from the cathode has become excited either when being released or when colliding in the gas phase. When the excitation is discharged, the atoms of the cathode material emit the characteristic emission spectrum. The emission of the lamp also includes the spectrum of the filler gas.

The advantages of the hollow cathode lamp include a relatively good intensity and the sharpness of the spectral lines [148]. Lamps are available for nearly all the chemical elements that can be determined by an atomic absorption spectrometer. A drawback is that every material to be determined needs a lamp of its own and the serviceable life of the lamps is, moreover, limited. Hollow cathode lamps are also available as so-called multi-element lamps, the cathode material of which contains two to four elements.

Nowadays there are microwave-tuned electrodeless discharge lamps available for some elements (e.g. As, Se, Sn and Pb). These lamps also emit a sharp-lined spectrum, their intensity being three to five times higher than that of a normal hollow cathode lamp. They need a separate power source, however, which is relatively expensive.

8.4.2. Atomisation of Sample

The most common way to atomise a sample is to conduct it into a high-temperature flame so that the chemical bonds of the sample are broken and free atoms are generated in the flame.

The sample solution is sucked into the flame usually through a capillary tube and a pneumatic nozzle, and in the mixing chamber of the burner there are normally flow barriers to separate out the larger drops, so that only an aerosol as homogeneous as possible is admitted into the flame. The most commonly used flame is the air–acetylene flame. By means of this flame it is possible to determine about 30 chemical elements. The temperature of the stoichiometric flame is 2450 K.

Some materials constitute oxides so strong that they do not decompose in the air–acetylene flame. They can be determined using an N_2O–acetylene flame. The temperature of the stoichiometric flame is 3200 K.

Sometimes also a hydrogen–air flame or a butane–air flame is used. The flame method has the disadvantage that the temperature of the flame cannot be controlled accurately.

   In the flameless method, a graphite tube heated by electric current is normally utilised as the atomising device, into which the sample is inserted by using a micropipette.

   In a graphite furnace the heating can be programmed. During the drying phase, the solvent evaporates. The thermal decomposition or the "ashing" of the sample is normally carried out at a temperature less than 1300 K. In the atomising step, the temperature is increased so high that the material to be determined is atomised. The highest temperature is 3300 K.

8.4.3. Detection Limit

It is thus possible in a graphite furnace to adjust the atomisation temperature to the most suitable for each material. In addition to this, the concentration of atoms in the absorption region grows much higher than in a flame, since only 2%–3% of the sample goes into the flame. This is why the detection limits for most materials, when using a graphite furnace, are 5–1000 times lower than when using the flame method. Table 8.3 shows examples of detection limits when using the graphite furnace and flame methods.

Table 8.3. Detection limits of atomic absorption spectrometry when using the graphite furnace and flame methods [148]

Element	Detection limit		
	Graphite furnace		Flame, relative $(\mu g \, l^{-1})$
	Absolute (g)	Relative $(20 \, \mu l^{-1}$ sample) $(\mu g \, l^{-1})$	
Ag	1×10^{-13}	0.005	2
Al	2×10^{-12}	0.1	20
Ba	5×10^{-11}	2.5	10
Cd	1×10^{-13}	0.005	2
Cr	1×10^{-11}	0.5	3
Cu	2×10^{-12}	0.1	1
Fe	3×10^{-12}	0.15	10
Pb	2×10^{-12}	0.1	10
V	1×10^{-10}	5	50
Zn	5×10^{-14}	0.0025	1

8.5. X-Ray Fluorescence

X-ray fluorescence analysis is a method of qualitative and quantitative analysis of chemical elements. Its detection limit is considerably higher than that achievable by means of the atomic absorption spectrometry, but is sufficient in many cases. Its advantages are, among other things, applicability for automation, short analysis time, and the fact that the sample material does not deteriorate or change when analysed. X-ray fluorescence produces a spectrum characteristic of each element, from which the element can be identified using large tables which have been prepared. Quantitative X-ray fluorescence analysis can be performed by comparing the intensities of characteristic peaks with those obtained from reference samples. However, X-ray fluorescence analysis cannot be performed for the lightest elements.

8.5.1 Generation of X-Rays

When an electron beam is directed toward a material, there is an interaction between the beam and the electrons of the material. The energy of the incident electrons is reduced and converted into X-rays. The wavelengths of this radiation form a continuous distribution, the shortest wavelength of which λ_{min} corresponds to the highest energy of the electrons. The shortest wavelength depends on the acceleration voltage as follows:

$$\lambda_{min} = \frac{hc}{eV} = \frac{1.241}{V} \times 10^{-6} \text{ m}$$

where h is Planck's constant, c is the velocity of electromagnetic radiation in a vacuum, e is the electron charge and V is the acceleration voltage in volts.

In general terms, the relation between wavelength (in nm) and energy (E in keV) is represented by the equation

$$\lambda = \frac{1.24}{E}$$

X-rays are generated in an X-ray tube (Fig. 8.7), between the anode and cathode of which there is an accelerating voltage. This voltage accelerates electrons released from the cathode by heating so that the electrons strike the anode, generating X-rays. The electron beam also makes the anode material heat up, and for that reason there is water cooling in the X-ray tube; without this the anode would rapidly melt. The form of the distribution of the intensity of the continuous radiation obtained from an X-ray tube is shown in Fig. 8.8. On top of the continuous distribution there are peaks of the radiation characteristic of the anode material of the tube. The intensity and its distribution is dependent on the operation voltage of the tube, the current and the material of the anode. X-ray fluorescence analysis uses this radiation to generate characteristic X-rays from the material studied. The spectrum of the exciting radiation must be continuous so that there are the necessary values of radiation energy for the excitation of each element. Fluorescence means that an electromagnetic radiation generates lower-energy radiation when it is incident on the atoms of the material.

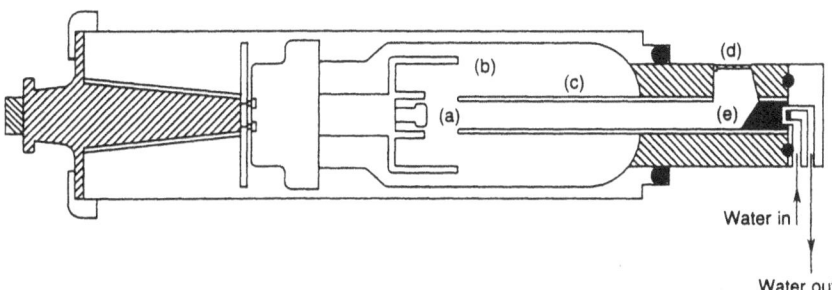

Figure 8.7. Cross-section of an X-ray tube [149]. The source of the electrons is a tungsten filament (a). The voltage between the cathode (a–b) and anode accelerates electrons through the focusing tube (c) onto the anode (d). A considerable part of the X-rays comes out from the window (e) of the tube.

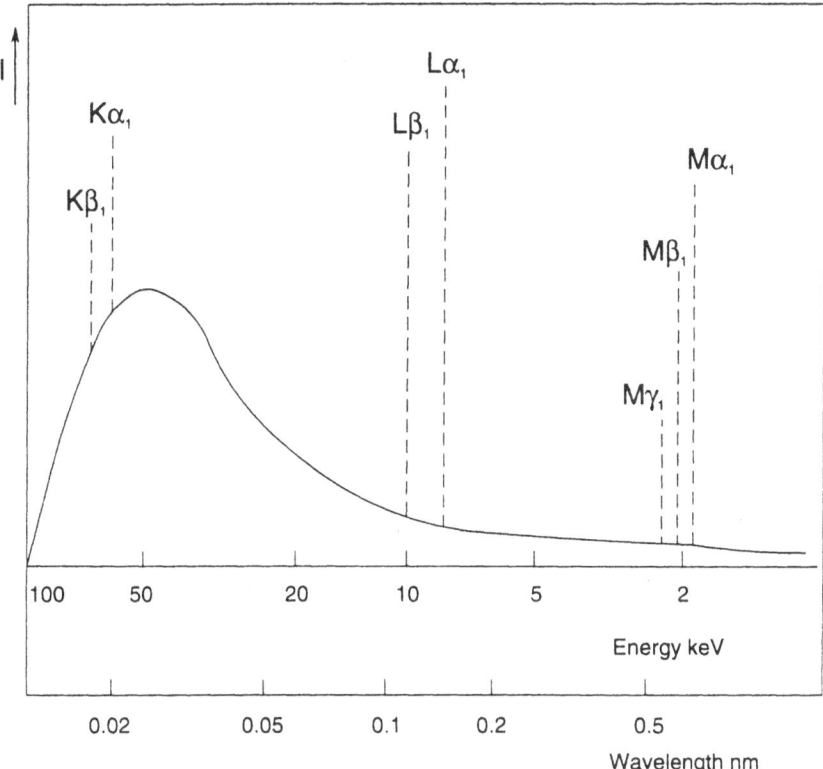

Figure 8.8. The intensity of the radiation obtained from an X-ray tube as a function energy and wavelength [149]. The anode material of the tube is tungsten and the voltage is 100 kV. The positions of the peaks of the characteristic radiation are also marked in the figure.

8.5.2. Fluorescence Analysis

The characteristic X-rays of the sample material in the X-ray fluorescence method are thus achieved by means of the primary radiation. It is directed to the sample from the X-ray tube. This radiation excites the energy states of the electrons in the sample material, for example, by removing an electron from the *K* shell of the sample material. To replace this electron a new one is immediately transferred from, for instance, the *L* shell, because this transfer means the discharge of the excited state, that is to say, the reduction of the total energy of the excited atom. The energy released in this transfer is removed away in the form of X-rays, the energy and thus wavelength of which is accurately determined by the difference in energy between the *K* and *L* shells of the atom in question. Since the *L* shell contains electrons, the energies of which differ slightly from each other, two characteristic radiations of the *K* series with different energies correspond to the transitions from the *L* shell to the *K* shell. These are denoted by $K_{\alpha 1}$ and $K_{\alpha 2}$. The peaks $L_{\alpha 1}$, $L_{\alpha 2}$, $L_{\beta 1}$, $L_{\beta 2}$ and so forth, proceeding towards transitions from higher electron shells, correspond to the electron transitions to the *L* shell.

The energy states of the electrons of an atom are typical for each atom. Therefore, there is an X-ray fluorescence spectrum characteristic of each atom, in which intensity peaks occur at the wavelengths and energies corresponding

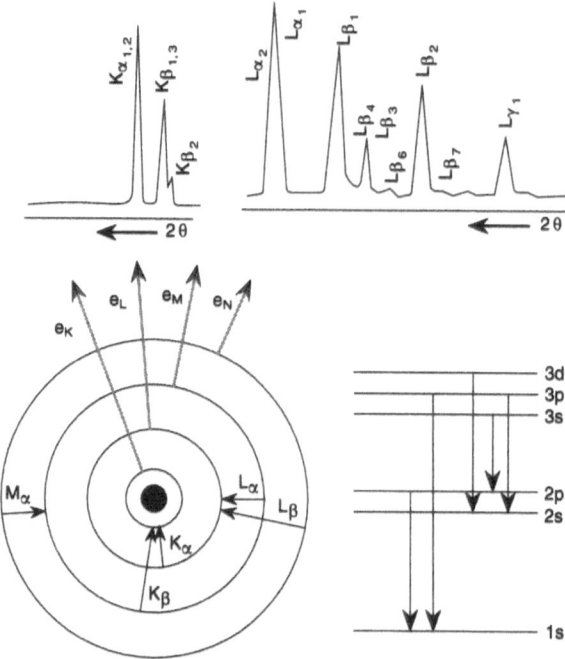

Figure 8.9. Generation of the characteristic X-rays and registered intensity peaks. For example, the electron transition between *K* and *L* shells generates the K_α peak.

to the transitions between electron shells. Figure 8.9 shows the generation of characteristic X-rays and the intensity peaks representing them. The qualitative X-ray fluorescence analysis is based on the identification of characteristic radiation spectra or peaks.

In the generation of the characteristic X-rays, the electron transfers thus take place between the inner electron shells of atoms. The energy differences are then great compared with, for instance, transitions among outer electrons, which can correspond to ultraviolet light and molecule fluorescence.

The intensity of the peak of the characteristic X-rays is proportional to the concentration of the element in the sample producing the peak. This is the basis of the quantitative X-ray fluorescence analysis. In order to realise it, reference samples are used, the concentrations of which are known. The intensity of the fluorescence peak obtained from the sample is compared with the intensity of the same peak obtained from the reference sample.

The detection limit of X-ray fluorescence analysis depends on the atomic number of the element analysed and on the sample matrix where it is located. The lightest elements are the most difficult to observe. The detection limit is further affected by the time used for the counting of pulses. For example, the detection limit of sodium in an alumina matrix is about 250 ppm when using a counting time of 100 s. The detection limit for sulphur in oil in similar conditions is of the order of 1 ppm.

Registration of X-Ray Spectrum

The registration of the X-ray fluorescence spectrum according to wavelength is carried out by means of a radiation detector and an analyser crystal, which are synchronously rotated. This is shown in Fig. 8.10. The X-ray spectrum is analysed as a function of the angle of rotation by means of the analyser crystal. This is a single crystal, and the distance between the various lattice planes is appropriate for the wavelength of the X-rays analysed. The operation of the analyser crystal is based on the Bragg law

$$n\lambda = 2d \sin \theta$$

where d is the distance between the diffracting planes, λ is the wavelength of the radiation, n is an integer and θ is the angle of diffraction.

The analyser crystal transmits only one wavelength when the diffraction takes place. By rotating the crystal and the detector so that the detector is rotated by the angle 2θ when the analyser crystal is rotated by the angle θ, the characteristic X-ray spectrum can be registered as a function of the angle and, according to the Bragg law, as a function of the wavelength. The device that turns the analyser crystal and the detector synchronously, and which is a part of the X-ray spectrometer, is called a goniometer.

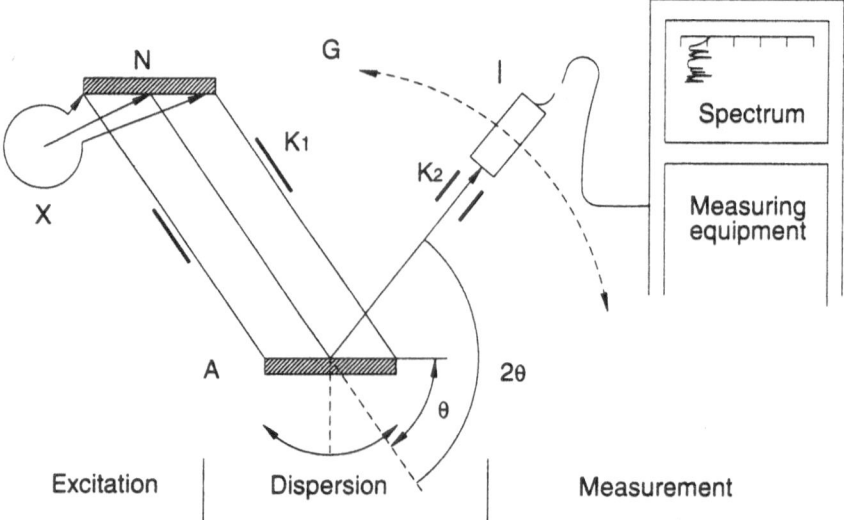

Figure 8.10. Schematic drawing of an X-ray spectrometer [149]. X, X-ray tube; N, sample; K_1 and K_2, collimators which make the X-rays parallel; A, analyser crystal; I, detector; G, goniometer.

Chapter 9

Economics of Emission Measurement and Control

9.1. Environmental Business

The world environmental market was estimated, in 1989, to be worth about US $277 billion and is projected to grow, by the year 2000, to US $454 [150]. The North American environmental market, for instance, could be summarised as follows:

In the US, the environmental market has already reached $130 billion and is growing about 7% per year. Approximately 2% of the US Gross National Product is spent on the environmental market and this is expected to grow.

The total Canadian market is estimated as being between $8.5 and 10 billion annually. In 1987, the environmental market of Ontario alone was estimated to be worth $1.5 to $2.5 billion and growing 17% annually. It is estimated, that the total Canadian market will exceed $20 billion by the year 2000.

In Mexico, the total pollution control sector, which is totally dependent on imports, can be divided into two distinct markets: instruments and equipment. In 1989, the instruments accounted for about 2% of the market and totalled US $5.3 million. In 1992 it was expected to reach US $7.4, all still imported. In 1989 the total market for equipment was US $212.4 million and it was expected to reach US $273 million in 1992. In the light of the Mexican developments and of the North American Free Trade Agreements, the Mexican environmental market is expected to grow dramatically.

Compared with the North American market, the European market is somewhat smaller. The market is over US $50 billion and is forecasted to grow well over US $100 billion by 2000 [150]. In Germany, for instance, more than US $25 billion was invested in environmental technology in 1990.

The development of environmental technologies originates from two principal factors. One of them is that clean technologies are considered to give advantages on the product market, and the other is that environmental investments open up new possibilities for the production and marketing of advanced environmental technologies.

9.1.1. Emission Measurements

Requirements for the environmental control create a basis for the investments in the measurement and analysis techniques. Emission limits have become stricter, and companies may have to pay large fines for violations of those limits.

Air pollution control is one of the basic objectives for the development of the environmental technologies. In the whole world, air pollution control was estimated to form about 25% of the total market of the environmental technology in 1990. Measurement and analysis instruments and process control systems were estimated to be the most rapidly growing areas.

In the USA, for instance, the market for measurement instruments related to emissions and air quality is estimated as shown in Table 9.1 [151], and the market according to the type of measurement instruments is shown in Table 9.2.

Table 9.1. Estimated market for measurement instruments in the USA related to emissions and air quality

	1991 (%)	1996 (%)
Emission monitoring	37	39
Occupational health and safety	33	25
Ambient air (outside)	16	14
Quality of inside air	14	22

Table 9.2. Estimated market for measurement instruments in the USA

	1991 (%)	1996 (%)
Emissions (CEM[a])	52	55
Portable instruments	25	27
Laboratory	23	18

[a] CEM means continuous emission monitoring

Authorities and enterprises will invest in environmental measurements in growing figures [152]. In the USA, the total market in 1989 was $1.16 billion. The market is expected to grow by 14% annually, and reach $2.25 billion by 1994. A large part of this ($1.5 billion) will be spent on laboratory instruments.

More regular emission measurements, including continuous monitoring, will be required by standards and authorities. In the Netherlands, for instance, large combustion units (over 300 MW) must have continuous measurement for SO_2, NO_x and particulate emissions [23]. In the USA, annual investments in "field" instruments are about $300 million for air quality instruments and $280 million for water quality control. The highest increase is foreseen in the instrumentation meant for the air quality control (21% annually). This is due to low initial level and to the general tendency and development of the legislation. In the USA, the highest growth figures (possibly more than 40%

annually) are expected for the measurement instrumentation for hazardous air pollutants as defined by EPA [153].

9.2. Cost–Efficiency of Sulphur Emission Control

The damage caused by atmospheric sulphur emissions to soil and lakes through acidification cannot always be most efficiently decreased by simply reducing emissions [154]. One should also take into acccount the atmospheric transport of emissions and the deposition patterns which they finally produce. The cost–efficiency of emission control measures with respect to deposition in a certain region has been calculated.

We can start the analysis by grouping the measures available for the reduction of emissions [154]. Let the number of these measures be $n(i)$. We know the unit cost of each measure, P ($ per ton SO_2), as well as the capacity V (ton SO_2) of that measure. By capacity we mean the maximum amount of emission to be reduced by the method at the unit price.

The cost–efficiency curve of the reduction measures for the emission source S_i can be constructed in the following way [154]: we sort out the measures in order according to the growing unit cost and define numbers $E_{i,j}$ and $C_{i,j}$, $j = 1, \ldots, n(i) + 1$, so that

$$E_{i,1} = \text{unreduced emission}$$
$$E_{i,j+1} = E_{i,j} - V_{i,j}, \quad 1 \leq j \leq n(i)$$
$$C_{i,1} = 0$$
$$C_{i,j+1} = C_{i,j} + V_{i,j} P_{i,j}, \quad 1 \leq j \leq n(i)$$

The cost $C_i(E_i)$, when $E_{i,n(i)+1} \leq E_i < E_{i,1}$, is now defined from the condition

$$C_i(E_i) = C_{i,j} + \frac{C_{i,j+1} - C_{i,j}}{E_{i,j+1} - E_{i,j}} (E_i - E_{i,j})$$
$$= C_{i,j} - P_{i,j}(E_i - E_{i,j})$$

when $E_{i,j+1} \leq E_i < E_{i,j}$.

In the resulting curve, there are then $n(i)$ linear segments, and in each segment j, the opposite number of the derivative, $P_{i,j}$, gives the limiting value of the cost of the emission reduction measures at the emissions corresponding to the segment. The cost–efficiency curve illustrates the lowest possible cumulative costs of the reductions at each emission level. The order of the measures obtained when constructing the curve gives the most favourable order of the application of the reduction methods [154].

When estimating the deposition of sulphur at a certain location, we must take the transport factors into account. This can be done by introducing a transport coefficient characterising the transport between the source and the receptor. This coefficient is normally based on models developed to describe the transport of emissions, such as the EMEP long-range transport model. The amount of background deposition originating from either natural sources or from unidentifiable human sources must also be judged.

The cost data should include the estimated costs of different emission reduction technologies as well as the costs of fuel switching in the case of energy plants. Furthermore, the costs for reducing process emissions from the major SO_2-emitting industrial sectors relevant to the areas concerned should be included.

The results of the calculations performed for some Finnish provinces indicate, that even if the most cost-effective measures can be taken in southern Finland, measures in Russian nearby regions should also be considered as having a high priority [154]. Significant reductions in deposition levels can, however, be achieved only through marked abatement of emissions in Central and Eastern Europe. An example of a cost–efficiency curve is given in Fig. 9.1. This curve illustrates the situation in south-eastern Finland.

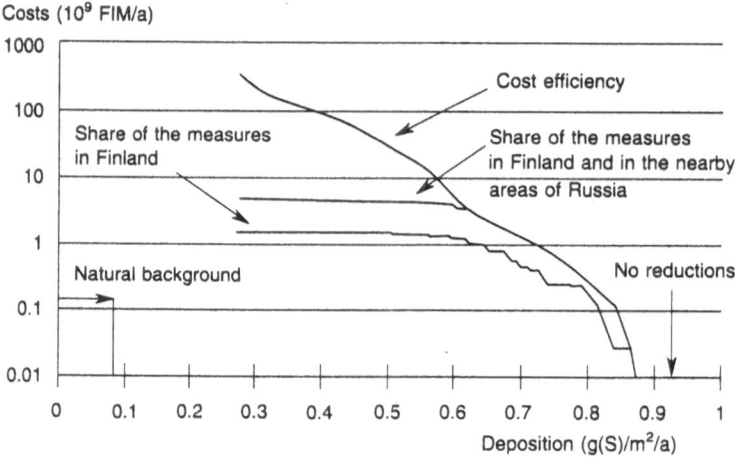

Figure 9.1. Cost–efficiency of the reduction of sulphur deposition in Southern Karelia, Finland. Emission reduction measures of the whole of Europe are included in the cost–efficiency curve [154].

9.3. Control of Nitrogen Oxides

About 95% of all NO_x from stationary combustion sources is emitted as NO [2]. Nitric oxide (NO) is formed by either or both of two mechanisms – *thermal* NO_x or *fuel* NO_x.

Thermal NO_x is formed by reactions between nitrogen and oxygen in air used for combustion. The rate of formation of thermal NO_x is extremely temperature sensitive and becomes rapid only at flame temperatures of 1600–

2000°C. Fuel NO_x results from the combustion of fuels that contain organic nitrogen in the fuel (primarily coal or heavy oil). Nitrogen bound in the fuel is highly reactive compared with the nitrogen contained in combustion air. Fuel NO_x formation is dependent on local combustion conditions (oxygen concentration and mixing patterns).

The two broad categories of NO_x controls are combustion modifications (primary methods) and flue gas treatment techniques (secondary methods) [2]. Combustion modifications are used to limit the formation of NO_x during the actual combustion. Flue gas treatment techniques are used to remove NO_x from flue gases after the NO_x has been formed.

There are several factors that contribute to high NO_x formation. Combustion controls reduce NO_x formation by one or more of the following strategies [2]:

1. Reduce peak temperatures of the flame zone.
2. Reduce gas residence time in the flame zone.
3. Reduce oxygen concentrations in the flame zone.

These changes to the combustion process can be achieved by either (a) modification of operation conditions on existing furnaces, or (b) purchase and installation of newly designed (low-NO_x) burners or furnaces.

Both process modification and new burner/furnace designs rely on the following concepts to implement the three main strategies for reducing NO_x emissions [2]:

1. Reduce peak temperatures by
   - using a fuel-rich primary flame zone.
   - increasing the rate of flame cooling.
   - decreasing the adiabatic flame temperature by dilution.
2. Reduce the gas residence time in the hottest part of the flame zone by
   - changing the shape of the flame zone.
   - using the steps listed in the preceding Strategy 1.
3. Reduce the O_2 content in the primary flame zone by
   - decreasing the overall excess air rates.
   - controlled mixing of fuel and air.
   - using a fuel-rich primary flame zone.

Intensive research and development efforts have led to a number of successful tactics that can be used to reduce NO_x formation without buying new burners or furnaces [2]. These tactics include

1. Low-excess-air firing
2. Off-stoichiometric combustion (includes overfire air)
3. Flue gas recirculation
4. Reduced air preheat
5. Reduced firing rates
6. Water injection

It has been estimated that additional investments in low-NO_x techniques in new power plants in Finland in 1987 have been in the range of 5000–11000 FIM/MW_{el} (electrical) or 1700–4200 FIM/MW_{fuel} (approximately US $1000 to 2200 or 350 to 900, respectively). These costs are due to new types of burners (staged combustion) [155]. For old power plants, when using primary methods, the estimates vary depending on the constructional features of the plants:

- Burner improvement   15 000–65 000 FIM MW_{el}^{-1}
  (normal to staged)   5600–12 500 FIM MW_{fuel}^{-1}
- Staged combustion and   65 000–100 000 FIM MW_{el}^{-1}
  flue gas recirculation   24 500–37 700 FIM MW_{fuel}^{-1}
- Combined staging of combustion   65 000–220 000 FIM MW_{el}^{-1}
  air and fuel, flue gas recirculation   24 500–83 000 FIM MW_{fuel}^{-1}

It has been found that the cost effectiveness for retrofit installations for the combustion modifications can be in the range of US $1500–3000 per ton NO_2 removed [156]. When stricter limits for NO_x emissions come into effect, more effective and much more expensive methods must be applied. Such technologies as selective noncatalytic reduction (SNCR) or selective catalytic reduction (SCR) systems and combined SO_x/NO_x removal processes can then be used as secondary reduction methods.

9.4. Taxes on Emissions

In order to be able control emissions on an economical basis, some countries have introduced taxes on emissions. In Sweden, for instance, taxes and charges concerning emissions of carbon dioxide (CO_2), sulphur (S) and nitrogen oxides (NO_x) have been decided. The CO_2 taxes in Sweden are added to the price of fuel and are imposed on all imported fossil fuels but not on peat or other domestic fuels like wood chips or solid wastes [157]. If coal, with a heat content of 7.6 MWh ton^{-1}, is used in a boiler with an average efficiency of 85%, the CO_2 tax will mean an added cost of almost 0.1 SEK kWh^{-1} useful heat. For fuel oils the CO_2 tax means around 0.08 SEK kWh^{-1} and for natural gas about 0.06 SEK kWh^{-1}. The legislation provides for a user rebate where the user has reduced CO_2 emissions.

The sulphur tax for coal and peat is 30 SEK kg^{-1} sulphur in the fuel [157]. For fuel oils with a sulphur content higher than 0.1% by weight the sulphur tax is 270 SEKm^{-3} and for each percentage of sulphur in the oil. For a heavy fuel oil with 1% S the sulphur tax means a cost of 0.03 SEK kWh^{-1} useful heat or around one-third of the CO_2 tax.

The sulphur tax has no impact on the heating costs for single family homes where oil with less than 0.1% S, natural gas, wood or electricity is used for heating. Coal and peat with high sulphur content will become increasingly less attractive. Wood and other domestic fuels will become more competitive as there is no sulphur tax on solid wastes. One of the problems with the sulphur tax is the need for methods of determining the real sulphur content in fuel.

A rebate of 30 SEK kg^{-1} S is paid if the emission of sulphur is reduced by cleaning or by absorption on ashes. The user has to prove how much sulphur is made harmless and there the analysis methods are very important. Technologies useful for reduction of sulphur oxides in flue gases are well established and owners of boilers fired with coal or heavy fuel oil with existing cleaning facilities in operation are now given a possibility to reduce operation costs through the rebate [157].

The NO$_x$ charge has been imposed from 1 January 1992 on all owners of "combustion units for energy production", meaning boilers or gas turbines with a capacity of 10 MW (fuel input) or more and useful energy output of at least 50 GWh per year. Soda recovery boilers in the pulp industry are excluded.

The NO$_x$ charge is 40 SEK kg^{-1} nitrogen oxides counted as NO$_2$. The quantity of emitted NO$_x$ will be measured and recorded with equipment approved by the authorities. If the emissions are not measured, the NO$_x$ charge will be based on an assumed NO$_x$ emission of 0.25 g MJ^{-1} fuel input for boilers and 0.6 g MJ^{-1} for gas turbines. The charge 40 SEK kg^{-1} and the emission 0.25 g MJ^{-1} give 0.01 SEK MJ^{-1} (fuel input) or 0.036 SEK kWh^{-1}. This value and an efficiency of 80% lead to a NO$_x$ charge of around 0.045 SEK kWh^{-1} or 2.25 MSEK per year for a boiler producing 50 GWh per year, the minimum cost for energy production per unit for which NO$_x$ charge will be imposed [157].

The NO charge differs in nature from the CO$_2$ and S taxes. All the money raised from the NO$_x$ charges, excluding the administrative costs, are proportionately paid back. There is a net expense for a boiler with above average NO$_x$ emissions and a net income for a boiler with emissions lower than average value. The charge period is over the calendar year. The owner has to deliver a yearly report for the actual energy production units covering amount of fuel used in the unit, amount of useful energy and the total NO$_x$ emitted. All data must be based on results of measurements carried out with equipment already approved by the authority.

It can be assumed that the NO$_x$ emissions per unit of energy where the resulting charge is zero, will slowly decrease, but as long as the charge is kept constant, 40 SEK is saved for every kilogram of NO$_x$ not emitted.

As a conclusion based on the Swedish emission tax and charge practice it can be said that the carbon dioxide tax is the most important of the three emission taxes. It makes the price of heat from coal, fuel oil and natural gas 0.1–0.06 SEK kWh^{-1} more expensive [157]. As no reduction methods for CO$_2$ are available, the only way to decrease additional costs is to try to increase the overall efficiency.

The sulphur tax is only placed on fuels with considerable sulphur content, i.e. some qualities of coal and heavy fuel oil. The sulphur emission can be reduced by using well-developed cleaning technologies which, even if they are not free of charge, are available and the emission level is more an economical issue than a technical one [157]. The sulphur tax has no impact at all on fuel costs for single family homes. On heavy oil with 1% S the sulphur tax causes an additional cost which is about one-third of that caused by the CO_2 tax.

The NO_2 charge is not a tax in the same sense as the others and most of the income will be recycled. Even if the average emission goes down to $0.15\,g\,MJ^{-1}$, the additional cost will be less than $0.02\,SEK\,kWh^{-1}$ for the owner who decides not to take any measures at all [157].

9.4.1. Carbon Dioxide Taxes

In the UN Meeting on Environment and Development in Rio de Janeiro in 1992 the United Nations Framework Convention on Climate Change was signed by 153 countries. This agreement is aimed at stabilising atmospheric concentrations of greenhouse gases at a level that would prevent harmful influence on the climate [158].

The practical realisation of the Rio agreement will need much consideration. The stabilisation of the emissions of carbon dioxide at the present level will require active economical control to be conducted in industrialised countries. Many countries should introduce a tax of about US $100 per ton carbon dioxide by 2000 [158]. By 2020 this tax should be US $150 and by 2050, US $300–500 per ton carbon dioxide. A tax of US $100 means an increase of 150% in the price of coal, and an increase of 60% in the price of oil. Consequences for the world economy would not, however, be intolerable. It has been estimated that their effect on the OECD countries, for instance, would be 2%–4% of the national product developed by 2050.

9.5. Reduction of Emissions Based on Measurements

The practical realisation of the air protection legislation requires monitoring of the emissions. Reliable data on emissions will also have a great economic significance as emission taxes and reduction technologies are applied. The need for emission measurements will then become considerably greater.

In addition to the air pollution control technologies, emission measurements also form the basis for the development of technologies in order to achieve lower emission levels. Measurements must be performed including all significant process stages. After the emission analysis, steps towards the improvement of processes and process engineering to reduce emissions can be taken. The operation of the emission gas cleaners must also be continuously monitored.

References

1. Hupa M, Boström S, Nermes M. Energiantuotannon kokonaispäästöt Suomessa. Kauppa- ja teollisuusministeriö, energiaosasto, Sarja D:162, Helsinki 1988
2. Cooper CD, Alley FC. Air pollution control: a design approach. Copyright © 1986 by C. David Cooper and F. C. Alley. Reissued by Waveland Press Ill 1990
3. van der Kooij J, van der Brugghen FW. Environmental control measures in Dutch power stations, 14th congress of the world energy conference, "Energy for tomorrow", Montréal 17–22 September 1989, pp 1–16
4. Hupa M, Boström S. Fluidized bed combustion; prospects and role. Report 91-7, Åbo Akademi University, Department of Chemical Engineering, Combustion Chemistry Research Group, 1991
5. Euroopan ilma – Euroopan ympäristö, Pohjoismaiden ministerineuvoston raportti, toimittanut Britt Aniansson, Tukholma 1986
6. Ilmansuojelu Suomessa. Esite 14 Ympäristöministeriö, Ympäristönsuojeluosasto, 1989
7. Aittola JP, Remo S. Polyaromaattiset hiilivedyt ja raskasmetallit hiilen ja turpeen poltossa, Imatran Voima Oy, R&D-Notes IVO-B-06/89, 1989
8. Williams PT. The Sampling and analysis of dioxins and furans from combustion sources. J Inst Energy 1992; 65: 46–54
9. Dioxins in the environment: pollution paper No. 27, Department of the Environment, UK, HMSO 1989
10. Mroueh UM, Tirkkonen T, Laukkarinen A, Hoffrén H. Energiantuotannon päästöjen mittaaminen. PAH-, PCDD- ja PCDF-yhdisteiden näytteenotto ja analytiikka, Valtion teknillinen tutkimuskeskus, Tiedotteita 817. 1988
11. Toimialoittaisen ilmansuojelutyöryhmän mietintö, Komiteamietintö 1981:70, Helsinki 1981
12. Mroueh UM, Sarkkinen S. Haihtuvien orgaanisten yhdisteiden päästöjen rajoittamismahdollisuudet. Kemia-Kemi 1991; 18 (1): 11–17
13. Valli R. Liikenne ja ilmansuojelu. Ilmansuojelu-uutiset 1991; 13 (3): 5–8
14. Jantunen M, Nevanlinna L. Kasvihuoneilmiö, ilmastonmuutos ja Suomi. Gummerus, Jyväskylä, 1990
15. Firth P, Averner MM. Biospheric-atmospheric interactions: the biological basis of climate. In: Kainlauri E, Johansson A, Kurki-Suonio I, Geshwiler M (eds) Energy and environment 1991, American Society of Heating, Refrigeration and Air-Conditioning Engineers, Helsinki, 1991, pp 77–85
16. Ruuskanen J. Atmospheric physics and chemistry. Lecture notes
17. Soud HN. Emission standards handbook: air pollution standards for coal-fired plants. IEA Coal Research, London, 1991
18. Offen GR, Altman RF. Issues and trends in electrostatic precipitation technology for U.S. utilities, J Air Waste Manag Assoc 1991; 41: 222–227
19. Dhali SK, Sardja I. Dielectric-barrier discharge for processing of SO_2/NO_x. J Appl Phys 1991; 69: 6319–6324
20. Emmel TE, Maibodi M, Martinez JA. Comparison of West German and US flue-gas desulfurization and selective catalytic reduction costs. Report No. PB-90-206319/XAB, 1990, Radian Corp. Research Triangle Park, NC (USA)
21. McIlvaine RW. New developments in European power plant air pollution control. Power Engin 1989; February: 33–35

22. Ando J, Kaplan N. Recent developments in SO_2 and NO_x abatement technology in Japan and China. Proceedings of the First Combined Flue Gas Desulfurization and dry SO_2 Control Symposium, 25–28 October 1988, St. Louis, USA

23. Mettälä K, Supponen M. Katsaus Hollannin ympäristöteknologiaan. Teollisuussihteeri-raportti 3/1991, Teknologian Kehittämiskeskus, Helsinki 1991

24. Salvaderi L. The emission control: problems, measures, effects on the future ENEL generating system. Rev Énergie 1990; 367–376 (n° 422, juillet-août)

25. Makansi J. Will combined SO_2/NO_x processes find a niche in the market? Power 1990; September: 26–28

26. Strömberg L. NO_x Legislation in Scandinavia. Three Day Topic Oriented Technical Meeting "In-furnace NO_x reduction techniques" (TOTeM3), 6–8 November 1990, Leeds University, UK. International Flame Research Foundation

27. EVA 4100. Electronic velocity array. Kurz Instruments Inc., 1986

28. Hahkala M. Future requirements for process and emissions diagnostics in conventional combustion power plant. Paper presented at the Finnish–French colloquium on combustion and emissions diagnostics, Tampere, Finland, 22 January 1993

29. Entropy Environmentalists, Inc. Preparation and review of site-specific test plans. Guidebook. US EPA Contract No. 68D90055, EMB Work Assignment No. 2-98, 1991

30. Niskanen V. Kaasumittaukset. Insinööritieto Oy, Espoo, 1985

31. Vahlman T. Kalibroinnin ja näytteenoton vaikutus kaasujen jatkuvatoimisessa päästömit-tauksessa. Ilmansuojelu-Uutiset 1990; 13(4): 11–14

32. Brandl A. Kalibrierung von Messgeräten zur laufenden Aufzeichnung von Emissionen, WLB Wasser, Luft Betr 1983; 30: 36–41

33. Koponen J. Kalibrointikaasut ja niiden käyttö. Kemia-Kemi 1986; 13(10): 849–855

34. SFS 3866. Ilmansuojelu. Päästöt. Kiintoaineen määritys manuaalisella menetelmällä; © Suomen Standardisoimisliitto, Helsinki, 1990

35. International Standard ISO 9096. Stationary source emissions—determination of concentration and mass flow rate of particulate material in gas-carrying ducts—manual gravimetric method. ISO, Geneve, 1992

36. Utrustring för mätning av stofthalter, STL-medi, Instruktioner, METLAB miljö Ab, Enköping, Sweden

37. Hesketh HE. Fine particles in gaseous media. Lewis Publishers, Chelsea, Michigan 1986

38. Environnement S.A. Emission particulate monitor model BETA 5 M, 1988

39. Ahonen PP, Kauppinen EI, Valmari ST, Mäkynen J, Joutsensaari J, Ylätalo SI, Lind TL, Jokiniemi JK. Continuous method for pulverized coal combustion emission mass monitoring. J Aerosol Sci 1992; 23, S1: S639–S642

40. Ruuskanen J. Optiset hiukkasmittaukset. Lecture notes INSKO 42–85, 1985

41. Read FH. Electromagnetic radiation. John Wiley, Chichester, 1980

42. Allen E. Colorant formulation and shading. In: Grum F, Bartleson CJ (eds). Optical radiation measurements. Vol. 2. Color measurement. Academic Press, New York, 1980, pp 289–336

43. Ruuskanen J. Studies of optical extinction for measuring mass concentration of spherical particle populations. PhD Thesis, University of Joensuu, 1984

44. Huovilainen RT. Pölypitoisuuden mittaus laserilla. Lecture notes INSKO 37–89, 1989

45. EMICON-laserpölymittari, Asennus- ja käyttöohjeet

46. Centre d'Etudes et Recherches des Charbonnages de France, Proces-verbal d'evaluation, pulverimetre à laser. Oldham-France 1983

47. Hinds WC. Aerosol technology. Properties, behavior and measurement of airborne particles. John Wiley, New York, 1982

48. Noll KE, Davis WT, Duncan JR. Air pollution control and industrial energy production. Ann Arbor Science Publishers, Ann Arbor 1975

49. Marple VA, Willeke K. Inertial impactors. In: Lundgren DA et al. (eds). Aerosol measurement. University Presses of Florida, Gainesville, 1979, pp. 90–107

50. Kuopion korkeakoulu, yhdyskuntahygienian laitos. Kahden jyrsinturvetta polttavan kattilalaitoksen haitta-ainepäästöt. Kauppa- ja teollisuusministeriö, energiaosasto, Sarja D: 41, Helsinki 1983

51. Vesterinen R. Energiantuotannon päästöjen mittaaminen. Näytteenottomenetelmät, mitta-laitteet, työtavat ja niiden kenttäkelpoisuus. Valtion teknillinen tutkimuskeskus, Tiedotteita 887, Espoo 1988

52. EPA Method 5. Determination of particulate emissions from stationary sources. 1986. Code of Federal Regulations 40. Parts 53 to 60 (40CFR53–60), pp. 503–524

53. Main IG. Vibrations and waves in physics. Cambridge University Press, Cambridge London New York Melbourne, 1978
54. Sears FW. Optics. 3rd edn. Addison-Wesley, Reading, Mass. Palo Alto London, 1964
55. Brown EB, Modern optics. Reinhold, New York, 1965
56. Hirleman ED, Oechsle V, Chigier NA. Response characteristics of laser diffraction particle size analyzers: optical sample volume extent and lens effects. Opt Eng 1984; 23: 610–619
57. Swithenbank J, Beer JM, Taylor DS, Abbot D, McGreath GC. A laser diagnostic technique for the measurement of droplet and particle size distribution. In: Zinn BT (ed.) Experimental diagnostics in gas-phase combustion systems: AIAA Prog in Astronaut Aeronaut, 1977; AIAA 53: 421–447
58. Chin JH, Sliepcevich CM, Tribus M. Determination of particle size distributions in polydispersed systems by means of measurements of angular variation of intensity of forward-scattered light at very small angles. J Phys Chem 1955; 59: 845–848
59. Alger TW. Polydisperse-particle-size-distribution function determined from intensity profile of angularly scattered light. Appl Opt 1979; 18: 3494–3501
60. GIV Produktinformation. Verfahrenstechnik. Meteorologie. Staubmesstechnik. Vakuumtechnik. Magnetik. Kryotechnik. Gesellschaft für Innovative Verfahrenstechnik MBH, Breuberg
61. Heiskanen M, Kauppinen E, Hahkala M. Energiantuotannon päästöjen mittaaminen. Hiukkasten sähköinen luokittelija ja sen sovelluksia, VTT Tutkimuksia 512, Espoo 1987
62. Seinfeld JH. Atmospheric Chemistry and Physics of Air Pollution, John Wiley & Sons, New York, 1986
63. Keady PB, Quant FR, Sem GJ. Differential mobility particle sizer: a new instrument for high-resolution aerosol size distribution measurement below 1 μm. TSI Quart 1983; 9: 3–11
64. ten Brink H, Plomp A, Spoelstra H, van de Vate J. A high resolution electrical mobility aerosol spectrometer (MAS). J Aerosol Sci 1983; 14: 589–597
65. Liu B, Pui D. Electrical neutralization of aerosols, Aerosol Sci 1974; 5: 465–472
66. Liu B, Pui D. Equilibrium bipolar charge distribution of aerosols, J Colloid Interface Sci 1974; 49: 305–312
67. Knutson E, Whitby K. Aerosol classification by electric mobility: apparatus, theory and applications. J Aerosol Sci 1975; 6: 443–451
68. Agarwal JK, Sem GJ: Continuous flow, single-particle-counting condensation nucleus counter. J Aerosol Sci 1980; 11: 343–357
69. Condensation Nucleus Counter Model 3020. Instruction manual. 1984, TSI
70. Boström CÅ et al. Avskiljning av flyktiga ämnen i stoftavskiljare, EM 1765, Institutet für Vatten- och Luftvårdsforskning, Göteborg 1985
71. Larjava K, Kauppinen E. Raskasmetalliaerosolien mittaus savukaasuista. Valtion teknillinen tutkimuskeskus. Tutkimuksia 414, Espoo, 1986
72. Strobel HA. Chemical instrumentation: a systematic approach 2nd edn. Addison-Wesley, Reading, Mass. 1973
73. Chen RF. Extrinsic and intrinsic fluorescence in the study of protein structure: a review. In: Guilbault GG (ed.). Fluorescence, theory, instrumentation, and practice. Dekker, New York, 1967, pp 443–509
74. Kalliorinne K, Kankaanperä A, Kivinen A, Liukkonen S. Fysikaalinen kemia 1, Kvanttikemia ja spektroskopia. Kirjayhtymä, Helsinki, 1988
75. Schwarz FP, Okabe H. Fluorescence detection of sulphur dioxide in air at the parts per billion level. Analyt Chem 1974; 46: 1024–1028
76. Okabe H, Splitstone PL, Ball JJ. Ambient and source SO_2 detector based on a fluorescence method. J Air Pollut Contr Assoc 1973; 23: 514–516
77. AF 20 M UV Fluorescent sulfur dioxide analyzer, technical start-up and maintenance manual. Environnement SA, 1984
78. Ohjeet ilmanlaadun mittaamisesta ja mittaustulosten vertaamisesta ohjearvoihin. Ympäristöministeriö, Ympäristön- ja luonnonsuojeluosasto, Sarja B, 7/1986
79. Mini-FTIR Gas Analyzer. Temet Instruments Oy, 1993
80. Charpenet L, Charpentier C, Fondanèche P. La mesure des oxydes d'azote, Description d'un analyseur à chimiluminescence à une ou deux voies de mesure indépendantes. Analusis 1983; 11: 327–340
81. Berry RS, Rice SA, Ross J. Physical Chemistry. John Wiley, New York 1980
82. Torvela H. Typen oksidien mittaus ilmansuojelussa. Kemia-Kemi 1988; 15: 22–24
83. Monitor Labs, Inc. Instruction manual, nitrogen oxides analyzer Model 8840, 1984
84. Signal, Model 3000, Operation Manual 1988

85. IR 703 Gas analyzer, Operations manual. Altamont Technologies Inc. Livermore

86. Gas Filter Correlation CO analyzer (Thermo Electron)

87. Leybold-Heraeus, BINOS^R IR/VIS/UV Gas Analysis System 1986

88. Rosenberg HM. The solid state: an introduction to the physics of crystals for students of physics, materials science and engineering. Clarendon Press, Oxford, 1984

89. Kittel C. Introduction to solid state physics. Wiley, Brisbane 1986

90. Verdin A. Gas analysis instrumentation. Macmillan, London 1973

91. Kaasinen P. Päästö- ja prosessikaasujen mittaukset laimentavaa näytteenottojärjestelmää käyttäen

92. Pocyn AJ, Hesketh HE. A review of current sampling and analytical methods for assessing toxic and hazardous organic emissions from stationary sources. J Air Pollut Assoc 1985; 35: 54–60

93. Harris JC et al. Sampling and analysis methods for hazardous waste combustion. Washington 1984. Environmental Protection Agency, EPA 600/8–84–002

94. Nottrodt A et al. Emissionen von polychlorierten Dibenzodioxinen und polychlorierten Dibenzofuranen aus Abfallverbrennungsanlagen. Müll Abfall 1984; 16: 313–330

95. Johnson L, Merril R. Stack sampling for organic emissions. Toxicol Environ Chem 1983; 6: 109–126

96. Sydor R, Pietrzyk D. Comparison of porous copolymers and related adsorbents for stripping of low molecular weight compounds from a flowing air stream. Analyt Chem 1978; 50: 1842–1847

97. Campbell R, Lee M. Capillary column gas chromatographic determination of nitro polycyclic aromatic compounds in particulate extracts. Analyt Chem 1984; 56: 1026–1030

98. Grimmer G, Naujack KW, Schneider D. Profile Analysis of polycyclic aromatic hydrocarbons by glass capillary gas chromatography in atmospheric suspended particulate matter in the nanogram range collecting 10 m^3 of air. Fresenius Z Analyt Chem 1982; 311: 475–484

99. Lee T, Schuetzle D. Sampling, extraction and analysis of polycyclic aromatic hydrocarbons from internal combustion engines. In: A. Bjørseth (ed.) Handbook of polycyclic aromatic hydrocarbons, Marcel Dekker, New York 1983, pp. 27–94

100. Hagenmaier H, Brunner H, Haag R, Kraft M, Lutzke K. Problems associated with the measurement of PCDD and PCDF emissions from waste incineration plants. Waste Manag Res 1987; 5: 239–250

101. Kendall DN. Applied infrared spectroscopy. Reinhold Publishing Corporation, New York 1966

102. Lindgren I, Nilsson J, Beckman O, Karlsson E, Kivikas T. Fysik 3, Kvantfysik. Almqvist & Wiksell, Stockholm 1971

103. Wieboldt RC, Hohne BA, Isenhour TL. Functional group analysis of interferometric data from gas chromatography Fourier transform infrared spectroscopy. Appl Spectrosc 1980; 34: 7–14

104. Malinen J, Tapola M, Uimonen J. Teollisuuden tarpeenmukainen ilmanvaihdon ohjaus ja anturitekniikka. Valtion teknillinen tutkimuskeskus, Tiedotteita 838, Espoo 1988

105. Huotari J, Linna V, Martikainen P, Rantanen J. IR- ja massaspektrometriset sekä akustiset mittaukset poltossa. In: Hupa M, Matinlinna J (eds). LIEKKI, Polttotekniikan tutkimusohjelma, Vuosikirja 1989, pp 289–312

106. Rantanen J. FTIR-spektrometrin käyttömahdollisuudet savukaasujen jatkuvatoimisissa pitoisuusmittauksissa. MSc thesis, University of Jyväskylä, Institute of Physics, 1988

107. Cygnus 100, 3159/1086, Mattson Instruments Ltd. Madison

108. Mills IM. Lecture notes. FTIR summer school, Helsinki 21–25 August 1989

109. Grim, WM, Fateley WG, Grasselli JG. Introduction to dispersive and interferometric infrared spectroscopy. In: Theophanides T. (ed), Fourier Transform Infrared Spectroscopy. Industrial Chemical and Biochemical Applications, Reidel, Dordrecht, 1984, pp 25–42

110. Stenman F. Spektroskopia. In: Kivalo P, Antila AM, Sihvonen ML, Konschin H, Paasivirta J, Stenman F, Sundholm F. Instrumenttianalytiikka 7, Optinen ja magneettinen spektroskopia. Teknillisten Tieteiden Akatemia, Helsinki, 1984, pp 537–689

111. Savolahti P, Fourier-infrapunaspektri kaasu- ja nestekromatografian detektorina. Kemia-Kemi 1986; 13: 11–14

112. Michelson Series user's manual, Version 2.0, preliminary version, Bomem Inc., 1989

113. Spartz ML et al. Evaluation of a mobile FT-IR system for rapid VOC determination. American Environmental Laboratory, November 1989

114. Strang CR, Levine SP. The limits of detection for the monitoring of semiconductor manufacturing gas and vapor emissions by Fourier transform infrared (FTIR) spectroscopy. Am Ind Hyg Assoc J 1989; 50: 78–84

115. Hanst PL, Lefohn AS, Gay W Jr. Detection of atmospheric pollutants at parts-per-billion levels by infrared spectroscopy. Appl Spectrosc 1973; 27: 188–198

116. Strang CR, Levine SP, Herget WF. A preliminary evaluation of the Fourier transform infrared (FTIR) spectrometer as a quantitative air monitor for semiconductor manufacturing process emissions. Am Ind Hyg Assoc J 1989; 50: 70–77

117. Saarinen P, Kauppinen J. Multicomponent analysis of FT-IR spectra. Appl Spectrosc 1991; 45: 953–963

118. Setälä R. Ilmanlaadun mittaukset Yhdysvalloissa. Teollisuussihteeriraportti 3/1993. Teknologian kehittämiskeskus, Helsinki 1993

119. Analysis of gases with opsis technology. Version 1, 27 September 1988

120. Grant WB, Kagann RH, McClenny WA. Optical remote measurement of toxic gases. J Air Waste Manag Assoc 1992; 42: 18–30

121. Powell I. Lecture notes

122. Eddy DS. Physical principles of the zirconia exhaust gas sensor. IEEE Trans Vehic Technol 1974; VT-23: 125–128

123. Daniels F, Mathews JH, Williams JW, Bender P, Alberty RA. Experimental physical chemistry. 6th edition, McGraw-Hill, New York, 1962

124. Hagan Probe Type oxygen analyzer package. Rosemount Analytical Inc., 1979

125. Worrell WL, Liu QG. A New sulfur dioxide sensor using a novel two-phase solid-sulfate electrolyte. J Electroanal Chem 1984; 168: 355–362

126. Gauthier M, Bellemare R, Bélanger A. Progress in the development of solid-state sulfate detectors for sulfur oxides. J Electrochem Soc 1981; 128: 371–378

127. Akila R, Jacob KT. An SO_x ($x = 2, 3$) sensor using β-alumina/Na_2SO_4 couple. Sensors Actuators 1989; 16: 311–323

128. Tournier G, Pijolat C, Lalauze R. CO Detection in town environment. EUROSENSORS, Third conference on sensors and their applications, Cambridge, 22–24 September 1987 pp. 162–163, the Institute of Physics

129. Torvela H. Capteurs de gaz à semiconducteur céramique. Application à l'étude des processus de combustion et au contrôle de la pollution atmosphérique. Rev Gén Therm 1990; no. 341: 1–7

130. Torvela H, Romppainen P, Leppävuori S. Detection of CO levels in combustion gases by thick-film SnO_2 sensor. Sensors Actuators 1988; 14: 19–25

131. Torvela H, Harkoma A, Leppävuori S. Detection of the concentration of CO using SnO_2 gas sensors in combustion gases of different fuels. Sensors Actuators 1989; 17: 369–375

132. Torvela H, Huusko J, Lantto V. Reduction of the interference caused by NO and SO_2 in the CO response of Pd-catalysed SnO_2 combustion gas sensors. Sensors and Actuators 1991; B, 4: 479–484

133. Harkoma A, Torvela H, Romppainen P, Leppävuori S. Detection of CO levels by semiconductor gas sensors in combustion gases containing NO. Combust Sci Technol 1988; 62: 21–29

134. Romppainen P, Torvela H, Väänänen J, Leppävuori S. Effect of CH_4, SO_2 and NO on the CO response of an SnO_2-based thick film gas sensor in combustion gases. Sensors Actuators 1985; 8: 271–279

135. Morrison SR. Measurement of surface state energy levels of one-equivalent adsorbates on ZnO. Surface Sci 1971; 27: 586–604

136. Clifford PK. Mechanisms of gas detection by metal oxide surfaces. PhD Thesis, Carnegie-Mellon University, Pittsburgh, USA, 1981

137. Lantto V, Romppainen P. Response of some SnO_2 gas sensors to H_2S after quick cooling. J Electrochem Soc 1988; 135: 2550–2556

138. Venema A, Nieuwkoop E, Vellekoop MJ, Ghijsen WJ, Barendsz AW, Nieuwenhuizen MS. NO_2 gas-concentration measurement with a SAW-Chemosensor. IEEE Trans Ultrason Ferroelect Freq Contr 1987; UFFC-34: 148–155

139. Brailsford AD, Yussouff M, Logothetis EM. Theory of gas sensors. In: Technical digest of the fourth international meeting on chemical sensors, Waseda University International Conference Center, Tokyo, Japan, 13–17 September 1992, pp 160–163

140. Gardner JW. Electrical conduction in solid-state gas sensors. Sensors Actuators 1989; 18: 373–387

141. Brailsford AD, Logothetis EM. A steady-state diffusion model for solid-state gas sensors. Sensors Actuators 1985; 7: 39–67

142. Wang X, Yee S, Carey P. An integrated array of multiple thin film metal oxide sensors for quantification of individual components in organic vapor mixtures. In: Technical digest of the fourth international meeting on chemical sensors, Waseda University International Conference Center, Tokyo, Japan, 13–17 September 1992, pp 526–529

143. Davide F, Di Natale C, D'Amico A. Sensor arrays figures of merit: definitions and properties. In: Technical digest of the fourth international meeting on chemical sensors, Waseda University International Conference Center, Tokyo, Japan, 13–17 September 1992, pp 390–393

144. Virkki J. Kokonaishiilivetyjen ja kevyiden hiilivetyjen mittaus. Lecture Notes INSKO 101–88, 1988

145. Pataki L, Zapp E. Basic Analytical Chemistry. Pergamon Press, Oxford, 1980

146. HP 5965B FTIR Detector for Gas Chromatography. Hewlett Packard, 1989

147. Rose ME. Modern practice of gas chromatography/mass spectrometry. VG Monogr 1990; 1: 1–23

148. Minkkinen P. Analyyttinen kemia. Lecture notes. Lappeenranta University of Technology, 1990, pp 49–67

149. Visapää A. Röntgenfluoresenssispektrometria In: Kivalo P, Sihvonen ML, Kuusi J, Pessa M, Visapää A. Instrumenttianalytiikka 5, Röntgen- ja fotoelektronispektrometria II. Teknillisten Tieteiden Akatemia 1981, pp 13–247

150. Sirkkola E, Porento L, Corbu O. Strategic alliances in environmental industry. Canada–Finland joint program 1989–1991. Teollisuussihteeriraportti 5/1992. Teknologian Kehittämiskeskus, Helsinki, 1992

151. Survey on Air Pollution instrumentation. Future Technology Surveys, Inc., Lilburn GA 1992

152. The environmental market: diversity and growth. R&D Magaz 1991; 18 February: 10

153. Air pollution monitoring market to experience surge in growth. J Air Waste Manag Assoc 1992; 42: 1506

154. Lehtilä A, Savolainen I, Tuovinen JP. Rikkilaskeuman kustannuksiltaan edullisin pienentäminen kotimaisin ja ulkomaisin päästönrajoitustoimin. Valtion teknillinen tutkimuskeskus, Tiedotteita 1212, Espoo 1991

155. Jaanu K. Energiantuotannon typpipäästöjen vähentämisen mahdollisuudet ja kustannukset. Kauppa- ja teollisuusministeriö, energiaosasto, Sarja D: 127, Helsinki 1987

156. Lepikkö J. Review of IVO's experience in NO$_x$ reduction techniques. Three day topic oriented technical meeting "In-furnace NO$_x$ reduction techniques", 6–8 November 1990, Leeds University, UK

157. Bäckström B. Taxes on Emissions in Sweden. Caddet Newslett 1991; no. 1: 2–3

158. Wallin M. Ilmastosopimukselle sinetti Riossa. Ilmansuojelu-Uutiset 1993; 16: 4–6

Subject Index